Landscape and Race in the United States

Edited by Richard H. Schein

Routledge is an imprint of the
Taylor & Francis Group, an informa business

Published in 2006 by
Routledge
Taylor & Francis Group
270 Madison Avenue
New York, NY 10016

Published in Great Britain by
Routledge
Taylor & Francis Group
2 Park Square
Milton Park, Abingdon
Oxon OX14 4RN

Printed in the United States of America on acid-free paper
10 9 8 7 6 5 4 3 2 1

International Standard Book Number-10: 0-415-94994-7 (Hardcover) 0-415-94995-5 (Softcover)
International Standard Book Number-13: 978-0-415-94994-1 (Hardcover) 978-0-415-94995-8 (Softcover)

Library of Congress Cataloging-in-Publication Data

Catalog record is available from the Library of Congress

Taylor & Francis Group
is the Academic Division of Informa plc.

Visit the Taylor & Francis Web site at
http://www.taylorandfrancis.com

and the Routledge Web site at
http://www.routledge-ny.com

Contents

CHAPTER 1

Race and Landscape in the United States

RICHARD H. SCHEIN

Introduction

This book on race and landscape in the United States begins with a simple image.[1] Figure 1.1 presents a portion of a 1903 fire insurance map.[2] Fire insurance maps became popular with insurance underwriters in the nineteenth century as a way to write policies without having to visit a site—they record building sizes, construction materials, proximity to water lines and fire hydrants, potentially flammable and combustible materials, and so on. They also carry a wealth of information that is useful beyond fire insurance underwriting—information that is factual even as it might shed light on American social and cultural attitudes extending well beyond a concern for fire in urban settings. This particular map depicts a small portion of the small village of Midway, in central Kentucky. It more specifically shows my house and yard as central, several other adjoining properties, my garage, and an ice house (ICE HO.) that was torn down long ago to make way for a new driveway. I start with my own house for two reasons. The first reason is prosaic, to make the point that this book intends to get the reader thinking about race and the cultural landscape in everyday places, such as one's own backyard. The second reason follows from the first but makes a somewhat broader claim that it is *always* possible to think about race and the American cultural landscape, *even* in one's own backyard.

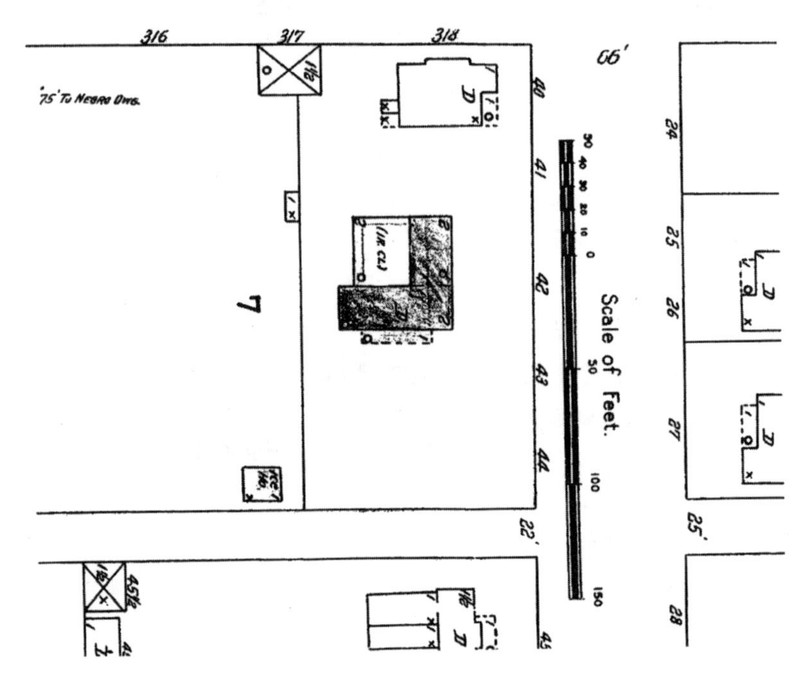

Figure 1.1 "Negro Dwg," or African American residences, are off the map, as indicated by a notation in the margins of this 1903 map from central Kentucky.

The argument begins with the words printed in the map's upper-left-hand corner: "75' to Negro Dwg" (Seventy-five feet to Negro dwellings). Clearly these words mark the presence of "race" in the landscape. Midway is in Woodford County, the majority of whose population on the eve of the Civil War had been African American, and the vast majority of them had been slaves in this rich, southern agricultural region. Even at the turn of the twentieth century, many of those slaves' descendants remained, and many of them had moved to town and lived, in 1903, in several small frame dwelling houses just "off the map" to the east, houses that sat next to a railroad line connecting Midway with the county seat. You might imagine this picture of race and landscape—the small African American neighborhood at the edge of (a southern) town was a regular feature of the Reconstruction South, even as that nascent urban residential pattern was solidifying through the aegis of Jim Crow into the kinds of segregated urban settings we can take for granted today. Midway was no exception. By the mid-twentieth century, there were "black" parts of town and "white" parts of town, a distinction that extended even to the one-block-long retail district running along both sides of the railroad track in the middle of town. We can perhaps imagine this scene, even as there is no map evidence (in 1903) of its wholeness—for the so-called Negro dwellings, the residences of African

Americans in Midway, were not mapped. I suspect the immediately utilitarian explanation for this absence, marked only by marginal notation, might be that African Americans were not expected to buy insurance. But this only deflects the larger questions of power in the economy (as in why weren't African Americans expected to buy insurance?) and in the landscape (as in why is it necessary to mark "Negro dwellings" even as they are absent from the map proper?). Those questions lead to interrogating the intersection of race and landscape in everyday American places, the first of my reasons for invoking the example. And they entail a certain bringing of race back onto the map—that is, if by race we mean the presence of Negro dwellings. And we often do, for "race" generally is treated as a marker for "people of color." It is a long-standing practice in American life that those deemed to have race are the ones who seemingly do not fit within the (white) majority.[3] Although it certainly is important to bring back to the map the presence of those small, African American neighborhoods as a central feature of Midway's landscape at the turn of the twentieth century, I also want to go a step further and do so through the claim that race already is present on the map, in other, not so obvious ways, even before we note the proximity of "Negro Dwg."

My house predates the Civil War. I know from the deed record and will books in the local county courthouse that the fellow who "built" the house in 1849 owned at least eight slaves, whom he passed on to his wife after his death shortly after the house was built.[4] I suspect that the house was built with slave labor, and I know that there were slaves living in and around the property, certainly during the antebellum period, but also free African Americans lived there well into the twentieth century. A little digging might uncover a wage–labor relationship between the people who used to live in my house and the people who used to live behind it. Here the seemingly common cultural landscape of a small town residential street also is fraught with race—only this time it is the construction of whiteness: construction in the material sense, where the luxury of white privilege in at least one small southern town was built, in part, on the backs of African Americans present, and construction in the metaphorical or theoretical sense in that our very ideas about (the normality of) whiteness depend on, in this case, what Toni Morrison calls an Africanist presence. "Seventy-five feet to Negro dwellings" looks like a cartographic manifestation of Morrison's "dark, abiding, signing Africanist presence." She has challenged a view of the American literary canon as "free of, uninformed, and unshaped by the four-hundred-year-old presence of, first, Africans and then African-Americans in the United States." Instead, she supposes an Africanist presence that "shaped the body politic, the Constitution, and the entire history of the culture ... the contemplation of this black presence is central to any understanding our national literature and should not be permitted to hover at the margins of the literary imagination." To explore this Africanist presence, how it functioned, and what it was for is also to ask, "What parts do the invention and development of whiteness play in the construction of what is loosely described as 'American'?"[5] In short, the whiteness of the map's central scene depended on the race of the off-the-map African Americans.

To bring those African American dwellings back onto the map is to interrogate the place of race in this landscape and is a twofold process, at least. The first part requires the simple acknowledgment that there were African Americans in that place in 1903 and that the full and proper story of the town and its landscape is not complete without their presence. This is conceptually akin to Stuart Hall's plea for the cultural and political need to recover hidden histories (and geographies, I would add) through "the speaking of a past which previously had no language."[6] Hall calls for "a struggle of the margins to come into representation" that seems relevant to the marginalized African Americans on the 1903 map. The second part requires us to realize that even parts of the town *not* usually associated with race, the *seemingly* normal landscapes of house and yard, are already coded "white" (in this case). And "whiteness" also is about race. That whiteness, however, is largely (and historically) invisible—at least to the hegemonic readings of race and landscape that presume white to be normal and everything else to be racialized.[7] In this sense, then, and following from Toni Morrison, *all* American landscapes can be seen through a lens of race, *all* American landscapes are racialized.

Now, up to this point, the example I have employed is firmly grounded in the dynamics of a black–white binary. Partly this stems from the place of the example in the American South, where white–black social relations have determined the social fabric for so long. Without denying the centrality of an Africanist presence in American life, the always racially coded spaces and landscapes of everyday life in America also extend to other formulations of race in its popular forms; that is, beyond a black–white binary to take seriously racial dynamics across the social spectrum. W.E.B. Du Bois once proclaimed that "the problem of the twentieth century is the problem of the color-line," and the point to follow here is that in the twenty-first century we are more apt to speak of color *lines*. But the underlying lesson is the same, and it also is the salient point of the following collection. The essays contained here are meant to get readers to think about the imbrications of race and the American landscape, even as they realize that all American landscapes are imbricated in questions of race.

To make that claim, however, requires a brief discussion of the key terms assumed so far: *landscape* and *race*.[8]

Cultural Landscapes and Race

The cultural landscape is both an object of study and a topic for any number of scholarly disciplines and professional practitioners. It is not the point of this volume to debate what Paul Groth and Chris Wilson called "the polyphony of cultural landscape study" or what Donald Meinig called "an attractive, important, and ambiguous term," even as it is necessary to have some idea of what the cultural landscape is and how it is that we think about cultural landscapes.[9] The term as it is most often employed in this volume comes from the tradition of landscape interpretation embedded in the discipline of cultural geography, while also drawing from other sources including anthropology, art history, and landscape architecture.[10] The term has historically connoted a prospect or a view upon the built environment, as well as its spatial ordering and material fabric.[11] The cultural landscape thus is a material thing, even as it invokes a way of knowing the world, an epistemology that relies in large part on vision. Peirce Lewis wrote of the cultural landscape that it is "our unwitting autobiography, reflecting our tastes, our values, our aspirations, and even our fears, in tangible, visible form."[12] Because of its qualities as a tangible, visible scene/seen, it follows that not only can we interrogate the historical and geographical dimensions of the landscape as an object in and of itself (as a material thing, or set of things), we also can read and interpret cultural landscapes for what they might tell us more broadly about social worlds of the past. In addition, cultural landscapes are not simply just *there* as material evidence in the service of observations about human activity. Their very presence, as both material "things" and conceptual framings of the world, makes cultural landscapes constitutive of the processes that created them in the first place—whether through the materiality of the tangible, visible scene or through the symbolic qualities they embed that make them inescapably normative.[13] Cultural landscapes are not innocent, and the duplicity of cultural landscapes means that we can, at once, study cultural landscapes as material artifacts, with traceable and documentable empirical histories and geographies, and simultaneously use cultural landscapes to ask questions about societal ideas about and ideals of, in this case, race in American life. We also can interrogate cultural landscapes as constitutive elements of those ideas and ideals. The ultimate goal of the essays in this book is maintaining that tension—between the landscape as a thing and the landscape as an entrée into ideas and ideals, realizing that the two rely on each other. Each of the essays asks the reader to think about a particular cultural landscape, the ideas and ideals it embodies, and, importantly, the tension inherent in the fact that those ideas and ideals, to some extent, rely on the landscape to give them life.

The essays in this volume cohere around ideas about and ideals of race. *Race* is a complicated and disputed term, with a long history and geography of use and misuse.[14] The essays in this volume take the concept of race, as well as racial categories, to be socially and politically constructed. The chapters draw generally from critical race theory an antiessentialist conception of race that nevertheless recognizes, following Cornel West, that "race matters"; that is, we act *as if* race is an ontological given.[15] This book begins by taking as given the "enduring role that race plays … in the very geography of American life."[16] Following Stuart Hall's call to correct the absences of past representation, this might entail simply noting the intersection of race and landscape in descriptive terms—a description of, say, the plantation in American antebellum life, or the historical importance of racial codes in keeping "order" in public space, or the role of the "New Negro Movement" in promoting African American business districts and racial pride, or the "gentleman's agreements" that were put in place to counter the "threat" of Chinatowns in the American West long before the immigration clampdowns of the 1920s. Such descriptions always run the risk of simply reinforcing the very categories and racisms that undergird them. But the stories of life and landscape that relied on racial differentiations were brutally real to many Americans and so deserve to be told. The fear of reifying racial categories is tempered by this volume's ultimate aim to focus on "racialized landscapes," a concept that draws on Omi and Winant's notion of racial formation, seeking to understand "the sociohistorical process by which racial categories are created, inhabited, transformed, and destroyed."[17] Racial processes take place and racial categories get made, in part, through cultural landscapes. Thus this book also takes as given the enduring role that the very geography of American life plays in understanding race. This book ultimately aims to draw attention to processes of racialization that take place through racialized landscapes.

That is a tall order, of course, and this book represents but a first step in the direction of understanding racialized landscapes. It is a first step that entails, at the very least, identifying and describing the intersections of race and landscape through a variety of empirical foci, through a variety of racial categories, including African American, Creole, white, Asian, Hispanic, and so on. Even as these categories are invoked, it is realized that they are not—even in their social construction or political utility—monolithic, which is to say that to identify someone or to self-identify as white or African American or Chinese is not to succumb to the ecological fallacy, or to assume that all people placed in or identifying with a category speak with one voice, or to presume adherence to a predetermined set of traits and behaviors traceable to the racial distinction employed. In addition, the book's antiessentialist stance as well as its ultimate aim to explore processes of racialization obviates the need to distinguish between *race* and *ethnicity*, even though it is possible to trace the differential uses and applications of the terms. It also is important to realize that identity is multiply constituted and that race is not the only axis of social power in people's lives or embedded in and working through everyday cultural landscapes. Gender and class, for example, also are powerfully ever present, and this book's *primary* focus on race is not meant to preclude other dimensions of identity formation, landscape interpretation, or social analysis. Indeed, individual authors in the following chapters directly engage these other axes of social power as they intersect with race and processes of racialization.

In short, the book is intended to help us *see* race and landscape in the United States, as a long-standing and key historical–geographical tension that is nevertheless often elided, perhaps because it is such an ordinary, everyday part of American life. Of course people who have been racialized generally need no lessons in seeing race in any aspect of American life, and for them the essays in this book may serve merely as vindication.[18] For others, however, I suspect that thinking critically about race and landscape might not come naturally, and these essays are aimed at prodding such analyses.

Seeing Race and Landscape in the United States

It is easy to see race in some landscapes. Birmingham's Kelly Ingram Park clearly engages race and racism and makes no bones about it. As West Park it was an organizing site for the Southern Christian Leadership Conference. It is linked to Dr. Martin Luther King's arrest that led to his letter from the Birmingham Jail. It is adjacent to the 16th Street Baptist church, where a Ku Klux Klan bomb killed Addie Mae Collins, Denise McNair, Carole Robertson, and Cynthia Wesley. It is the place were Bull Connor ordered the arrest of, by some accounts, more than one thousand children during civil rights demonstrations in the early 1960s. It is a key site in the American cultural imagination, and it is linked forever to the 1964 Civil Rights Act. A historic photograph of a young protester forced into the jaws of a police dog presumably was the inspiration for the contemporary memorialization of these events, represented in Figure 1.2. This cultural landscape is a powerful reminder of the place of race in American society. It is impossible to walk through this site without being moved. Indeed, for

Figure 1.2 It is easy to see race and landscape in Birmingham's Kelly Ingram Park, a powerful reminder of the civil rights movement and racism in American life.

some it is impossible to walk through the site. And although it purports to memorialize one heroic chapter of resistance, resilience, and perseverance in a brutal past, our engagement with the tangible, visible scene in Kelly Ingram Park does the work of bringing that past continually into the present.

Kelly Ingram Park is a *designed* landscape. It is consciously intended to make us think about race.[19] The task always is harder in the everyday, taken-for-granted scenes in the *vernacular* landscape, especially when the work of race often is seemingly invisible, hidden, or even overwritten in the landscape's palimpsest appearance. The courthouse square in Lexington, Kentucky, for instance, displays a couple of hundred years of civic and cultural monuments. Although the monuments and layout of the square suggest a design, it is one achieved through accretion, and the various monuments around the square result from discrete decisions made over the years by individuals and groups who built on a constantly evolving scene but who never have been guided by an overarching design imperative. The invisibility of race is striking on this site, once you recall that is was the nineteenth-century site of the public market known as Cheapside, which served as a slave market. Central Kentucky had a surplus slave population by the 1840s, and people were sold down the Ohio River to work in the westward-migrating Cotton Belt. Until recently, there was no historical marker to that effect, despite the very meaningful place of Cheapside in the geographical imaginations of many African American Lexingtonians (who have avoided the site and its surrounding businesses for generations). That invisible or even erased narrative also is at work in the design of the Kentucky quarter. The coin depicts a horse grazing behind a four-board fence in a scene redolent of Central Kentucky's Inner Bluegrass. The big house on the quarter is Federal Hill, ostensibly the place where Stephen Foster penned the minstrel ballad that is the state song. "My Old Kentucky Home" is Foster's idea of the lament of one of those slaves sold down the river, who recalls fondly life left behind in dear old Kentucky where, as the song's original second line tells us, " 'tis summer, the darkies are gay." In both of these scenes is an Africanist presence and an absent narrative only recently and reluctantly brought to the surface in Lexington's downtown through the enterprise of three African American brothers who finally enabled a historical marker to be emplaced in the courthouse square. In doing so, they brought the stories embedded in the site to light and insisted on a conscious attention to the cultural landscape of slavery and its legacies.

That continual presence often is linked to the structural imperatives of race and racism, the kinds of historical and geographical vestiges of race that seem to live on in American society beyond the control or imagination of any single individual. Often this is the place of cultural landscapes in processes of racialization. Many of us would honestly deny a racist intent in our daily activities, but the very structures of the world that we live in can make us unconsciously complicit in perpetuating processes of racialization through our interaction with and through the landscape. For example, the *New York Times* recently reported on the persistence of race restrictive deed covenants in American land practice, despite their contemporary illegality.[20] Deed covenants still are an important part of America's suburban landscape, related to planned developments, the role of home owners' associations, and niche marketing.[21] Racial covenants became popular as a way to restrict, especially, African American residents in America's burgeoning twentieth-century suburbs after racially restrictive zoning was declared illegal by the Supreme Court in 1917. Although the practice of *racial* covenants has been outlawed, many deeds still carry the restrictions in print and remind us of past racist practices that live on in the whiteness persisting in many once racially covenanted suburban communities. As Evan McKenzie suggested, "While the covenants are there, there is still room for people to think that although it cannot be legally enforced it is nonetheless a promise that they are morally obligated to keep."[22] That moral obligation may extend to the also illegal practices of redlining, where lending organizations declare a certain area of the city off-limits to mortgage lending, and real estate steering, where individual agents "lead" clients to one or another part of a city based on racial evaluations of a client. Most banks and real estate agents decry such practices, of course, but they do persist in some quarters, and the imagery of the American suburb still to this day is overwhelmingly one of whiteness.

The cultural landscape is especially adept at masking its complicity with processes of racialization when it is enacted as part of other, seemingly more benign narratives of American life. Historic preservation, for example, generally appeals to a broad segment of the population through its reliance on a landscape or architectural *aesthetic*, which is usually invoked as something beyond assail, as a cultural value that is not somehow tainted by the political. But the aesthetic is never simply common sense, and it is a learned appreciation that privileges particular ways of looking at, knowing, and valuing landscapes and architecture. Invoking the aesthetic, through something such as historic preservation, always uses the cultural landscape in ways that have consequences, including racialized ones, beyond the intent of those involved in any particular preservation effort.[23] But to move in that direction regarding race and landscape is to ask questions about how landscapes *work* in reproducing everyday life and all of its social relations. That is the place this book hopes to end. Rather than draw conclusions, I close this section with two juxtaposed images (Figure 1.3), separated by a couple of thousand miles and one hundred years yet both grounded in the American experience. The first is a highway sign in Southern California, near the U.S. border with Mexico; the other is a public monument in Holland, Michigan. These two images might speak volumes about attitudes toward race and ethnicity in the United States during, at least, the past two centuries. The image on the top warns you to look out for fleeing immigrants, and the allusion is to illegality, stealth, haste, and maybe even fear. The one on the bottom captures the stalwart nineteenth-century forebears of the upper Midwest, immigrants from two centuries past who presumably formed the backbone of today's democratic society and who are celebrated for their sacrifice and perseverance. The message is mixed: immigrants bad, immigrants good. And the message is fraught with racial connotations, simmering just below the surface of an American melting pot ideology.

Now, I know that last paragraph was rather glib and that the exercise really isn't fair: one of these images is about events taking place today, and the other carries within it the romantic gloss of historicized hindsight, a story written in the landscape by the third generation and sanctioned by HRH Princess Margaret of the Netherlands, who unveiled the

Figure 1.3 This figure features two representations from two different centuries of immigration in American life.

statue in 1997. You might argue that the image of running pedestrians is demographically wrong, and we could test that empirically in any given migration stream or city. And I might argue that the family-centered heteronormativity captured in the man–woman–child imagery tells more about the sign makers than those depicted and might be more historically accurate for the bundled northern immigrants than for those on the yellow highway sign. And you might suggest, politely, that these are just my interpretations, and you would be correct to point out that we cannot understand either of these images without, first, recourse to their specific and particular cultural contexts and, second, a sense of how they fit more broadly into the American experience writ large. And that, I suggest, is a good thing. Our ensuing conversation—even disagreement—is actually the point. That is what representations do; that is what symbols are for. Each of these images, to greater or lesser extent, carries the weight of American ideas about and ideals of race and ethnicity; these images are representations of American concepts and practices, and if we interpret them as symbolic of those American ideas and ideals, they also mediate our ongoing formulations of the very concepts we purport to marshal in our arguments. This, too, is the place of the cultural landscape, of racialized landscapes in American life. And that is what the essays in this book do—they begin with particular instances of race and landscape in America, they begin with description, they move to interpretation, and they intend to serve as a starting point, to get us seeing the landscape around us, and to get us talking about the topic of race in American life.

The Essays

The claim at the beginning of this chapter that race always is in the landscape precludes a comprehensive or totalizing view of race and landscape in the United States. In other words, there is no point in even trying to introduce the spectrum of American racial practice as context for the chapters that follow. Clearly there are key moments and events and landscapes in an American historical geography of race and landscape: slavery and the Thirteenth Amendment, Chinese exclusion laws, the Gadsden Purchase, sharecropping, redlining, the immigration reform acts of 1921 and 1924 and 1965, the "Gentleman's Agreement" with Japan in 1907, Jim Crow, the Ku Klux Klan, the New Negro Movement, historically black colleges and universities (HBCUs), the Harlem Renaissance, and urban renewal, among many others. Thus beyond the rather general foundational claims about race and landscape in the preceding section, this chapter makes no attempt to *a priori* set the stage for the essays following. Instead, each author addresses within his or her chapter relevant facts and contexts of American land and life. The selection of authors in this volume is, at first glance, idiosyncratic. I drew specifically on scholars of the American landscape who have been writing about race and landscape in a way that theoretically and thematically shares the intention of this volume. My own intellectual home in cultural geography has meant that most of the authors here are geographers, although attempts were made, some successfully, to enlist others beyond the discipline. The list of contributors also was limited to those who had both the time and the inclination to contribute, and the several obvious omissions generally reflect busy schedules and other projects rather than a conscious attempt to exclude. I deliberately moved beyond the black–white binary in choosing chapter subjects to include other racialized landscapes, including those identifiable as Latino or Hispanic, Chinese American, native Mexican, and white. Regrettably, many other significant intersections of race and landscape are missing, including those involving Native Americans. Also, I sought analyses of race in both urban and rural landscapes, and I tried for historical and contemporary examples and essays that link the historical with the contemporary. Last, I enlisted authors who would take seriously the imbrications of class and gender with race. So, although anything like complete coverage is perhaps impossible, perhaps the reader will grant some forgiveness for any particular omission and can see how the examples herein, although specific and selective, share similarities with other processes of racialization and other associated landscapes. If we are successful in this volume then our omissions will prompt additions to the fledgling literature on race and landscape in the United States, contributions to the ongoing conversation and argument about the interrelations of race and landscape.

In asking each author to write about race and landscape in the United States, I followed a modest formula designed to move from landscape description to landscape as a reflection of cultural practice to the concept of racialized landscapes, fully implicated in the ongoing production and reproduction of American life. The formula structures the charge to authors that I reproduce below. That formula suggests that it works to first take on the question of landscape history, such as when and where was the landscape created, by whom, why, how has it been altered, and so on. We then can start to ask questions about what the cultural landscape means by attempting an interpretation of the landscape as unwitting autobiography. And as we argue over what it means—about authorial intentions and readers' interpretations, over different interpretations—our conversations will range from the landscape to link with other big ideas in American life. At some point in this process, we will see the landscape as the tangible, visible node between everyday practice and ideas and ideals about, in this case, race, and we realize that we are engaged in interrogating the symbolic dimensions of landscape, asking how the landscape works, as discourse materialized, to normalize and naturalize social and cultural practice, to challenge it, and so on. In every case, however, the authors were asked to maintain the focus on the material form of landscape, the tangible, visible scene/seen, even as the landscape interpretation transcended that scene to range across any number of issues central to American life.

Each author was given the following, more specific charge:

1. Describe the landscape you are writing about. Include carefully selected images (including maps where relevant), not as "window dressing" but as central to introducing the landscape in question. The primary focus of this chapter should be on the tangible, visible scene/seen, on the landscape's materiality, and you might open with a vignette that lays that scene before the reader as the chapter's touchstone.
2. Undertake a *landscape history* of the landscape. What are its origins; who or what (laws? hegemonic practice?) was, has been, or is responsible for its creation, maintenance, destruction (this is about power and agency and authorship, and the place of such in broader societal structures)? Include an attention to the place of this landscape as either unique or of a type (including an understanding of why that is important).
3. What does or might your landscape *mean* to the people who live or lived in and through that landscape? To those of us who know it only from a distance? (Or link the landscape as part of everyday life to broader American ideas about society, economy, polity, and so on.)
4. Speculate on the importance of the landscape in question to American ideas about and ideals of race.
5. Do not be constrained by these charges so that you do not also interrogate other aspects of the landscape as central to questions of race in American life.

Of course, authors are not automatons, and so individual chapters vary from this formula, but in the end most of the following essays are structured in this way. Following, then, are eleven individual essays that collectively represent a contribution to thinking about race and landscape in the United States.

In the next chapter Michael Crutcher presents a two-hundred-year historical geography of Faubourg Tremé, just outside the original ramparts of New Orleans, and describes and documents the evolution of that neighborhood's landscape with regard to Creole, Free People of Color, and African American occupants. There is scope across the United States for such historical geographies, to simply understand the basic patterns and processes of those parts of American towns and cities that were historically identified by their residents' race and to realize that embedded in the often segregated landscapes of race are the tangible, visible markers of community, resistance, resilience, and identity formation central to the stories of identity, pride, and growth, as well as indications of, oftentimes, abandonment, renewal, tension, and conflict as older, African American parts of the city become attractive to other American residents in the familiar processes of gentrification. Crutcher ends with a coda commenting on the fate of New Orleans after Hurricane Katrina, as that city and the United States struggle to come to terms with issues of race and residence in American cities.

In chapter 3 Steven Hoelscher explores how landscapes of race and memory stand at the symbolic and political economic center of a struggle over questions of southern regional identity. Hoelscher introduces the southern plantation landscape as a nationally resonant symbol and discusses the invention of that landscape tradition after the Civil War, bringing it into the present day through the particular case of Natchez, Mississippi. He presents the semiannual Natchez Pilgrimage as an elite sponsored cultural practice that includes touring the most prized antebellum mansions in the city. The landscape of the "white-pillared past" is explored as one of the sites through which stories and rituals of citizenship are enacted and resisted, in this case through contemporary black counternarratives.

In chapter 4 Samuel Dennis also interrogates the plantation landscape through the example of Hampton Plantation, a State Historic Site and a stop on the South Carolina heritage landscape tour near Charleston. Dennis proposes that there are at least two very different "ways of seeing" the plantation landscape. The first predominates and entails a "planter's view" of the land linked to private property and social status and works to materially and discursively erase anything other than the white, owner's presence on the land. Through a long-standing Hampton Plantation owner's love of its land and life, we are introduced to the dominant narrative of planter paternalism that serves to naturalize race and gender categories in the postbellum plantation landscape. As a challenge to the cultural amnesia of that nostalgic view, one that often elides slavery and other racialized social relations, Dennis raises the possibilities and problematics of alternative narratives predicated on a second view, which specifically recounts historic links between African Americans—as both slaves and free citizens—and the land.

Chinese immigration through Angel Island in San Francisco is Gareth Hoskins's subject in chapter 5. Angel Island is a federal immigration station turned California State Park, and Hoskins presents it as a racialized landscape in two ways. First, between 1910 and 1940, Angel Island materialized exclusion, shutting out a presumed "Chinese menace" and had the effect of legitimating whiteness as an American norm. Second, as a contemporary historic site, Angel Island works in a more positive way to recount and mediate American debates about citizenship, race, and immigration. The chapter sets the immigration station's creation and transformations within a larger historical geographical context of American immigration policy and practices of memory, with particular emphasis on the views of those who passed through the site, including those whose voices are forever memorialized in poems inscribed on the barracks' walls. Angel Island is presented as a site where once excluded people consciously use the past to reclaim histories and territories in ongoing American dialogues about race and identity.

In chapter 6 Daniel Arreola analyzes the representation of Mexican housescapes in the southwestern United States in popular historic postcards. He documents how picture postcards captured, promoted, normalized, and ultimately created the Mexican housescape as a symbolic landscape that carried the weight of Mexican American misrepresentation. Postcards reinforced domestic U.S. stereotypes about Mexican Americans and as such are related to early twentieth-century concerns about the so-called Mexican problem. Images of supposedly typical Mexican American landscapes captured both a particular way of seeing and a substantive content that made them anything but value-neutral images of southwestern land and life. Arreola shows how they served to reinforce specifically pejorative images of Mexican Americans through a visual culture that still lives in popular representations of Mexican American, Latino, Latina, and Hispanic people in the contemporary United States.

Dianne Harris directs our attention to the landscape of suburbia through her detailed examination of the postwar American house in chapter 7. She presents ordinary houses as central to American racial, class, and ethnic assimilation during the time of burgeoning suburban expansion in the United States. Domestic building and design industries as well as government-sponsored lending practices and the actions of real estate developers and agents worked in concert to produce houses for a specific image of the "good American" who was, in this period, characterized by an unquestioned whiteness. Landscapes of home are seen as central to forming American cultural identities—through concerns with privacy, individuality, and racial conformity. Harris specifically examines the manner in which design professionals, including popular home magazines, as well as the very materials of construction were implicated in processes of postwar American class distinction and racialization.

The subject of chapter 8 also is suburbia, through the particular examination of two affluent, suburban landscapes. James Duncan and Nancy Duncan explore the adjacent New York suburbs of Bedford and Mount Kisco, which are linked materially through labor and conceptually through the concept of a landscape aesthetic. Bedford is presented as a widely supported and maintained rural, idyllic landscape, whose aesthetic both masks and implicates the town in the structures of white privilege. That landscape is maintained in large part by the Latino labor force, which arrives in Bedford daily from next-door Mount Kisco, a town that is struggling with its own aesthetic of public space in the face of a (dominant white) perceived "Latino invasion." These two landscapes of home are presented as politically contested symbols of personal, community, and regional identity and are not interrogated as isolated or even insulated spaces of white privilege but presented within a global–local matrix in which ideas about and ideals of "the aesthetic" work to reinscribe a dominant set of globalized and racialized social relations. The importance of their study site transcends suburban New York through claims about the importance of these particular suburbs as home to a class of decision makers whose influence extends beyond their local abodes.

James Rojas describes in chapter 9 how Latino immigrants and Mexican American citizens are retrofitting the built environment of East Los Angeles to meet their cultural needs. The rich practices of public life and ideas about home life and its relation to the street are transforming the East Los Angeles landscape spatially and visually. Rojas includes street vendors, fences, *la yarda* (or the enclosed front yard), parks and open space, an urban farm, and the politics of urban space (through the Latino Urban Forum) as he documents the evolution of what he calls an East Los Angeles vernacular landscape. The very presence of these landscape transformations speaks to the anachronistic qualities of much of East Los Angeles's contemporary landscape, to the need that people have to shape their everyday environments to meet their functional needs and their cultural predilections, and to the ever-changing nature of cultural landscapes, both as material things and through their place in constructing meaning and identity in peoples' everyday lives.

In chapter 10 Jonathan Leib discusses the controversy in Richmond, Virginia, surrounding a proposal to place a statue of Arthur Ashe—native, tennis star, social activist, and philanthropist—on that city's Monument Avenue, home to statues traditionally dedicated to heroes of the Confederacy. Richmond was the capital of the Confederate States of America, and the proposal was not taken lightly. Leib describes the history of Monument Avenue before taking on the 1995 debate about placing the statue of Ashe there. The whiteness of Monument Avenue and the city of Richmond ultimately were challenged (and not for the first time). The Arthur Ashe controversy presents a classic case where the cultural landscape mediates a set of debates that is at once about landscape and visuality even as it also transcends the cultural landscape to stand for race and racialized social relations in general, in Richmond, and, by extension, in American life.

Derek Alderman also engages memorializing landscapes, in particular the naming of streets after Dr. Martin Luther King Jr. In chapter 11 Alderman positions these streets as sites of struggle over cultural and political power, over claims to the city, over who has voice as urban citizen. Alderman traces the origins of streets named after King to black community activism and presents a national picture of the practice before interpreting the symbolism of the practice and of the streets themselves. He suggests that commemorative street naming not only invokes the memory of Dr. King but also serves to mediate debates about race and racism in American life generally. Many specific examples are presented to highlight the associated social, political, and economic tensions of naming streets after King that emerge through public debate and controversy. He concludes that naming streets after King symbolizes African American empowerment and struggle through the contemporary urban cultural landscape.

In the last chapter Heidi Nast seems at first to move far from the concerns of race and landscape in her description and interpretation of three (out of eleven at last count) pet parks in Chicago: Doggie Beach, Wiggly Field, and Puptown. The parks generally are located, however, in Chicago's gentrified, and largely white, north side and so are implicated in the social relations of urban commodification (of pets and their associated products) and the racialized social relations of the contemporary city (which displays a racialized urban historical geography). Nast's chapter is a fitting conclusion to this volume. Her essay clearly illustrates the point that opened this introduction—that not only is race visible in ordinary everyday cultural landscapes such as one's own backyard but it is always possible to think about race and the American cultural landscapes, even in one's own backyard.

Acknowledgments

I am grateful to Dave McBride for his stoic optimism and patience in bringing this volume to fruition; to Sue Roberts for reading this chapter's penultimate draft and providing insightful and helpful comments and suggestions; and to Dick Gilbreath for his usual above-the-call-of-duty willingness to perform cartographic and illustrative magic on short notice.

Notes

1. The book's title accurately reflects the focus here on cultural landscapes of the *United States*. The awkwardness of the term *United States* as an adjective means that throughout this chapter the term *American* is used to mean, in this case, United States even as I realize the many other legitimate claims to *American* extant.

2. Sanborn Map Company, *Midway, Woodford County, Kentucky* (Sanborn Map Company, 1903); one map on three sheets. It is a map I am rather fond of at the moment, and I have used it to illustrate the imbrications of race and landscape in several different ways, in several different contexts. See Richard H. Schein, "Digging in Your Own Backyard," in *Archivaria* (forthcoming); and Richard H. Schein, "Acknowledging and Addressing Sites of Segregation," *Forum Journal* (National Trust for Historic Preservation) 19 (2005): 34–40.
3. And in this vein we often speak of "tolerance"—as if those with race somehow are a nuisance that needs to be tolerated rather than full-fledged citizens with equal standing in a society.
4. County Clerk's office, Versailles, Woodford County, Kentucky, *Will Books N:413 (will) and N:465 (appraisal)* of R.H. Davis (1850).
5. Toni Morrison, *Playing in the Dark* (Cambridge, MA: Harvard University Press, 1990), 4–8.
6. Stuart Hall, "The Local and the Global: Globalization and Ethnicity," in *Culture, Globalization, and the World-System: Contemporary Conditions for the Representation of Identity,* ed. Anthony King (Binghamton: State University of New York, 1991), 34–35.
7. This point is of course "old news" to anyone not of the white majority population. People of color certainly have been aware of the always racially coded spaces and landscapes of everyday life in America for as long as there has been everyday life in America.
8. The book also assumes a focus on cultural landscapes in the *United States,* a focus that is not meant as normative but rather utilitarian and reflects nothing more than the need to bound the book at some level, as well as the limits to my own claims for understanding race and landscape in that particular national context. Clearly race is at work in other cultural landscapes around the world.
9. Paul Groth and Chris Wilson, "The Polyphony of Cultural Landscape Study: An Introduction," in *Everyday America: Cultural Landscape Studies after J.B. Jackson,* ed. Paul Groth and Chris Wilson (Berkeley: California University Press, 2003), 1–22; and D.W. Meinig, "Introduction," in *The Interpretation of Ordinary Landscapes,* ed. D.W. Meinig (New York: Oxford University Press, 1979), 1–7.
10. For a very good introductory overview, with further references, see Denis Cosgrove, "Cultural Landscape," in *The Dictionary of Human Geography,* 4th ed., ed. R.J. Johnston et al. (Oxford: Blackwell, 2000), 138–41. Representative texts not produced by geographers include Barbara Bender, ed., *Landscape: Politics and Perspectives* (Oxford: Berg, 1993); W.J.T. Mitchell, ed., *Landscape and Power* (Chicago: University of Chicago Press, 1994); and Matthew Potteiger and Jamie Purinton, *Landscape Narratives* (New York: John Wiley & Sons, 1998).
11. Denis E. Cosgrove, *Social Formation and Symbolic Landscape* (Totowa: Barnes & Noble Books, 1984); and Richard H. Schein, "Representing Urban

America: Nineteenth-Century Views of Landscape, Space, and Power," *Society and Space (Environment and Planning D)* 11 (1993): 7–21.

12. Peirce Lewis, "Axioms for Reading the Landscape," in *The Interpretation of Ordinary Landscapes,* ed. D.W. Meinig (New York: Oxford University Press, 1979), 12.
13. Richard H. Schein, "The Place of Landscape: A Conceptual Framework for Interpreting an American Scene," *Annals of the Association of American Geographers* 87 (1997): 660–80; and Richard H. Schein, "Normative Dimensions of Landscape," in *Everyday America: Cultural Landscape Studies after J.B. Jackson,* ed. Paul Groth and Chris Wilson (Berkeley: University of California Press, 2003), 199–218.
14. The literature on race is, of course, voluminous, and this book makes no attempt to do it justice. For an introduction to the historical depth of racist ideas, see Emmanuel Chukwudi Eze, ed., *Race and the Enlightenment: A Reader* (Oxford: Blackwell, 1997); and George M. Frederickson, *Racism: A Short History* (Princeton, NJ: Princeton University Press, 2002). For an introduction to the ideas of Critical Race Theory, see Richard Delgado and Jean Stefanic, eds., *The Cutting Edge* (Philadelphia: Temple University Press, 1995); and Richard Delgado and Jean Stefanic, eds., *Critical White Studies* (Philadelphia: Temple University Press, 1997). For an introduction to some of the ways in which geographers, especially social and cultural geographers and those interested in the cultural landscape, have approached questions of race, see several journal articles and theme issues, including Catherine Nash, "Cultural Geography: Anti-Racist Geographies," *Progress in Human Geography* 27 (2003): 637–48; Linda Peake and Richard H. Schein, "Racing Geography into the New Millennium: Studies of 'Race' and North American Geographies," *Social and Cultural Geography* 1 (2000): 133–42 (this essay introduces a theme issue that includes seven accompanying articles); Richard H. Schein, "Introduction," *Professional Geographer* 54 (2002): 1–5 (this essay introduces a theme issue on race, racism, and geography that includes seven accompanying articles); and Audrey Kobayashi and Linda Peake, "Racism out of Place: Thoughts on Whiteness and an Antiracist Geography in the New Millennium," *Annals, Association of American Geographers* 90 (2000): 392–403.
15. Cornel West, *Race Matters* (New York: Vintage Books, 1994).
16. Michael Omi and Howard Winant, *Racial Formation in the United States,* 2nd ed. (New York: Routledge, 1994), vii.
17. Richard H. Schein, "Teaching 'Race' and the Cultural Landscape," *Journal of Geography* 98 (1999): 188–90; Omi and Winant, 55.
18. David R. Roediger, ed., *Black on White* (New York: Schocken Books, 1998); and bell hooks, *Killing Rage* (New York: Henry Holt, 1995).
19. There is a burgeoning literature that explores race and the designed landscape. See, for example, Craig E. Barton, ed., *Sites of Memory* (New York: Princeton Architectural Press, 2001); and Lesley Naa Norle Lokko, ed., *White Papers, Black Marks* (Minneapolis: University of Minnesota Press, 2000).

20. Motoko Rich, "Restrictive Covenants Stubbornly Stay on the Books," *New York Times,* April 21, 2005 (online at http://select.nytimes.com/gst/abstract.html?res=F40D12F93D550C728EDDAD0894DD404482).
21. Evan McKenzie, *Privatopia* (New Haven, CT: Yale University Press, 1994).
22. Quoted in Rich.
23. Schein, "Normative Dimensions;" James S. Duncan and Nancy G. Duncan, *Landscapes of Privilege* (New York: Routledge, 2004).

CHAPTER 2

Historical Geographies of Race in a New Orleans Afro-Creole Landscape

MICHAEL CRUTCHER

Stepping out of New Orleans's French Quarter toward the northeast, the lower French Quarter to be exact, one crosses Rampart Street and enters Faubourg Tremé (see Figure 2.1).[1] The quick crossing into Faubourg Tremé can be thought of as leaving one neighborhood for another; tourist for residential, affluent for poor, safe for dangerous, dangerous for deadly. In essence one crosses several boundaries and enters many landscapes.

Houses in both neighborhoods are small, as a rule,[2] and sit adjacent to the banquette, or sidewalk, having little or no front yard, and with the exception of Governor Nicholls Street, one of the oldest streets in the city, trees are sparse and small. Residential architecture in the French Quarter and Tremé is dominated by Creole cottages and variants of the shotgun house: single, double, raised, and camelback. In the first blocks across Rampart Street, the houses in the Faubourg Tremé are in good states of repair; many recently renovated and painted in either strikingly vivid or muted pastel colors. The social and economic characteristics of this section of Tremé vary, but residents are whiter and wealthier than those farther "back o' town," where houses appear less well kept, many are dilapidated, and some are burned out.[3] Vacant lots become more prevalent, as are houses and apartment buildings that vary from the traditional New Orleans architectural styles. The built landscape is dotted with corner stores, bars, and

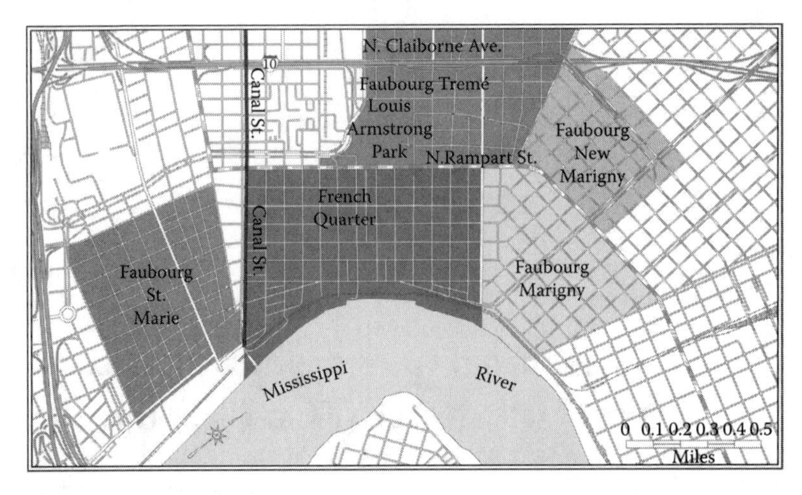

Figure 2.1 Central New Orleans: The French Quarter and Faubourgs Tremé, Marigny, New Marigny, and St. Marie.

occasionally more monumental churches and schools. This section of Tremé is home to low-income African Americans. Their lives are affected by the same social conditions that persist in similarly composed urban communities across the city, indeed the nation: low educational attainment, violent crime, and drugs. Louis Armstrong Park is here. It is a thirty-acre public space containing an accidentally postmodern assemblage of monumental edifices and historic spaces and buildings, including Congo Square, a nineteenth-century slave and Indian marketplace. Congo Square's importance comes from the African religious and musical customs practiced there during slaves' free time. Other noted structures in the park include the Morris F.X. Jeff Municipal Auditorium, the Mahalia Jackson Theater for the Performing Arts, the Masonic lodge Perseverance Hall No. 4, and the Rabassa and Reiman houses. In the fenced and gated park, these structures are variously arranged around landscaped berms and a cement lagoon. Last, the neighborhood is bounded or bisected (depending on ideas of territory) by an interstate highway elevated above Claiborne Avenue by row after row of concrete piers. Just beyond the neighborhood, but very much in the vicinity, are the fabled vaults and tombs of St. Louis cemeteries #1 and #2.

This is the landscape of the Faubourg Tremé (see Figure 2.2). In truth, the physical space just described varies not so greatly from other historic neighborhoods in New Orleans. The shotgun house, in all its variations, is a predominant New Orleans house type. Creole cottages are also present in other downtown areas, and the same can be said of other structures and institutions present in Tremé. Tremé, however, is one of New Orleans's

Figure 2.2 A view of the cultural landscape in Faubourg Tremé.

most historically significant and presently perplexing neighborhoods. Tremé's proximity to the French Quarter and its affordable historic architecture have made the neighborhood one of the most ripe for gentrification. The long-term residents, however, are some of the most vocal and best organized in the city and have seen fit to protest this movement by calling for the salvation of black Tremé by limiting gentrification and its alien influences.

This chapter presents a brief historical geography of race and cultural landscape in Tremé, incorporating several of the neighborhood's changing sociospatial contexts, including its European colonial development, its racial composition, changes in the local and national political economies, and juxtaposition to the French Quarter. Faubourg Tremé's contemporary landscape has many authors and several origins and can be viewed through the temporal lens of two development eras. The first can be traced to the city's early geographic expansion, the territory's transfer to the United States, and the resulting tensions of a Creole enclave within an increasingly American city. The second era begins with early twentieth-century attempts to revitalize Tremé and continues today in the tensions between long-term residents who consciously and unconsciously draw on the neighborhood's richly variegated Afro-Creole historical and social geographies, newly arrived gentrifiers, and the city of New Orleans.

The First Era: From the Beginning

Between its founding by brothers Jean Baptiste and Pierre Le Moyne in 1718 to the Louisiana Purchase in 1803, New Orleans developed according to design by infilling the ramparts of the military city. The ramparts, New Orleans's "fortified" walls, never approximated the massive fortifications represented on colonial maps but consisted of raised earth mounds topped with wooden palisades. Symbolically, however, these walls marked the separation of the city and the hazards of the hostile natural environment and threatening Native Americans.

The prospect of developing a successful colony within the city walls proved difficult. In the early years the colony was temporarily abandoned on several occasions. During those times people went to live with neighboring Indian tribes.[4] Socially, uneven sex ratios between African slaves and whites facilitated race mixing. The practice of mixed-race marriages and concubinage had long been practiced in the New World. Along the Gulf Coast, relationships between native women and white men were often preferred by the men and advocated by the church. After the arrival of African slaves in 1719, African and Indian unions became common. But it was the mixing of the French and African to create the Creole social category that would come to most characterize New Orleans's social structure. The biracial offspring of these unions always held the bonded status of the mother, who was usually a slave. The system that produced this third class or racial category was known as placage. The name refers to the white male practice of selecting a mistress and placing her and their children in a small house were they could live comfortably.[5]

In this system white fathers often freed both the mother and the child, producing a class of people variously known *gens de couleur libre* (Free People of Color [FPC]) or Creoles of Color. In addition to being free, this group often inherited property and money and reaped benefits from their partial white status. During the course of New Orleans history, however, lack of equal status radicalized a segment of the free population.

Two events were practically relevant to the physical and numerical expansion of New Orleans during the French (1718–62 and 1800–03) and Spanish (1762–1800) colonial periods. First, as a consequence of two Spanish-era fires, much of the city's rebuilt architecture was transformed from wood to sturdier brick.[6] In the rear and lower sections of the Quarter, beyond the fire's reach, Creole cottages and shotgun houses dominated. Those house types eventually spilled over the city's makeshift ramparts to become the typical house types for the expanding city.

Second, the city's population doubled as a result of political events in the Caribbean.[7] The island of Saint-Domingue witnessed a revolt by its free mulatto planter class (1790) followed by a revolt of former slaves (1791). The latter resulted in an independent Haiti (1804). These conflicts spread Haitians across the New World. Eventually, many found their way to New Orleans. This group, numbering close to ten thousand, was equally divided between whites, African slaves, and FPC.[8] To accommodate these new immigrants, developers and the city of New Orleans established the faubourgs, or subdivisions, Marigny, New Marigny, and Tremé.[9] The culture and institutions of these Francophone people came to distinguish the landscapes of these early neighborhoods.

In 1798 Claude Tremé first began to sell lots in his plantation just outside the ramparts. As of 1803 the early subdivision of Tremé was apparent. Not until 1812, though, would the city surveyor Jacques Tanesse officially plan "the New Faubourg Tremé." Five years earlier, in 1807, a surveyor laid out Faubourg Marigny, making it New Orleans's first suburb. Joseph Pilie plotted Faubourg New Marigny, adjacent to Faubourg Tremé, in 1809.[10]

The development of New Orleans's subdivisions coincided with the process of Americanization following the Louisiana Purchase in 1803. This process involved much more than a simple switch in colonial administrations. The territorial exchanges between France and Spain in 1762 and 1800 had been largely seamless because the population remained French.[11] The Louisiana Purchase, however, fueled an enmity not seen before. Americans, arriving from the north in increasing numbers and with differing linguistic, legal, and religious traditions, posed a very real threat to the Creole ways of life. One of the rumored outcomes of Creole and American antagonisms was a prejudice that forced the newly arriving Americans to reside as a group upriver of Canal Street in the Faubourg St. Marie. The Americans did settle predominately "upriver" of Canal Street. Although the boundary between the two sectors of town was nowhere as absolute as portrayed in the mythology of New Orleans, the Upriver (uptown)–Downriver (downtown) division remains the city's most lasting geographic boundary.[12]

In addition to the Creole, who was despised for being lazy and immoral, Americans also disparaged the status of slaves and FPC. Creoles of Color had liberties approaching those of whites, and slaves had guaranteed rights under the Code Noir, both of which were unfamiliar to Americans. Slaves were able to roam city streets without their masters, who hired them out as laborers, and they were given time to congregate and tend garden plots. Slaves and FPC also worshipped under the same roof as whites, though in separate pews.

The race question was addressed early in the American period when laws were passed to suppress the black militia[13] and limit black mobility. In 1808, 1817, and 1822, the city passed legislation limiting slave dancing and Sunday congregations to Place Publique.[14]

American attempts to override the social and cultural traditions of New Orleans created a contentious atmosphere in the city. However, although powerful politically, Americans remained in the numerical minority, thus allowing the Creole population to resist changes. Relatedly, many Creoles owned taxable property, giving them voting rights, whereas most Americans were still propertyless. Creoles also teamed with foreign French immigrants to solidify their Gallic superiority. The straw that broke the Americans' backs in their dealings with the Creoles was the Creole-dominated city council's purposeful sabotage of the American Faubourg's wharf, thereby threatening its very livelihood.[15]

According to Tregle, the Americans parlayed rural Louisianans' anti-French sentiment to effect a split of New Orleans into three semiautonomous municipalities. The first municipality included the French Quarter and the Faubourg Tremé. The second municipality included the Faubourg St. Mary (formerly the Faubourg St. Marie) and other areas upriver of Canal Street and served the new American population. The third municipality, downriver from the first, was basically identical to the Creole Faubourg Marigny. The new municipalities controlled only "internal financial and economic affairs," whereas citywide only one mayor and a single police force remained. Again, the Canal Street boundary between the first and second municipalities energized the Creole and American division.

Although the creation of New Orleans's municipal system likely seemed a victory for downtown Creoles, or at least a reprieve from the American onslaught, the 1830s, 1840s, and 1850s were decades of declining influence for the French Quarter and Creole culture in general.[16] Most glaringly, the Creole sector fell behind in commercial wealth. Tregle noted that Creole failures in education translated into Creole absences from certain professions and the declining use of the French language. A Creole enclave could not stop the growing political, economic, and demographic progress of the Americans. The dialectical tensions between Creoles and Americans, uptown and downtown, began to produce landscape change in New Orleans in general and Faubourg Tremé specifically, especially through Creole religious practices and the changing status of African descended slaves and Creoles of Color.

In the first decades of Americanization, New Orleans's religious leaders attempted to realign St. Louis Cathedral's liberal religious tradition with the conservatism of southern planter society. For more than forty years, the church's French-speaking board of trustees successfully stonewalled attempts by the American bishops to install pastors sympathetic to American sentiments. Important, and in fact integral, to St. Louis Cathedral trustees' governance success was their prominence as French Freemasons. The trustees' egalitarian Masonic ideals contributed to the "strong current of religious ecumenism at St. Louis Cathedral."[17]

Upset that St. Louis Cathedral was being run by its lay board (known as Marguillers) composed of Masons, American Catholics took steps to instill more conservative clergy. Following a protracted battle, the cathedral was wrested from the hands of the Marguillers. For the city's black and white Catholics, the change signaled a betrayal by one of their most central cultural institutions; a betrayal that resulted not only in new allegiances but in changes in the cultural landscape.

The new racial dynamics of the cathedral were ameliorated by the construction of St. Augustine's church in Tremé in 1842. The church served Tremé's growing population of white Creoles, French immigrants, FPC, and slaves. It is accepted that half of St. Augustine's congregation were FPC. The church's construction was preceded by the formation of an order of predominantly FPC nuns in 1836. The order, eventually called the Sisters of the Holy Family, had the goal of educating the area's poor African-descended population. The order ran several schools and asylums in the Tremé area dedicated to this population.

Although St. Augustine's and the Sisters of the Holy Family were aligned with the concerns of New Orleans's and Tremé's FPC and slaves, many Creoles became disenchanted with the Catholic Church and began to seek out anticlerical organizations and affiliations, such as Masonic and benevolent societies that constructed meeting halls in Tremé that served multiple purposes over the years. A benevolent society example is the Société d'Écônomie d'Assistance Mutuelle (better known as the Economy Society), formed in 1834. In its early decades the society, composed of aristocratic FPC, sponsored charity balls at its hall on Tremé's Ursuline Avenue. The group also focused on highlighting the political causes of emancipation and suffrage for FPC. In the decades following the Civil War, the Economy Society possessed more property ($4,000) than any black society and organized the Economy Hotel Joint-Stock Company, one of New Orleans's most sound black joint-stock companies.[18] Although always sponsoring functions requiring music, in the twentieth century, Economy Society Hall became noted as a venue for early jazz music.

Although many organizations similar to the Economy Society flourished in the nineteenth century, it was in the years after the Civil War that the number of societies swelled. To survive in the difficult postbellum years, freedmen and FPC shunned charity, turning instead inward toward mutual aid and benevolent societies. Through dues, fines, and taxes, these groups were able to provide or subsidize medical care and burial expenses. Some of the organizations located in or associated with Tremé were La Société des Artistes (The Society of Artists) (1834), Société d'Écônomie (Economy Society) (1834), Les Jeunes Amis (The Young Friends) (1867), and La Concorde (1878). Most of these organizations built halls in or very near Tremé.[19]

The activities in Tremé following the Civil War continued to develop the neighborhood's landscape. Following the failure of radical reconstruction, a politically active group of Creoles of Color calculated ways to challenge efforts underway to solidify and codify black inferiority. Tremé Creoles organized the Comité des Citoyens in 1890 and brought to the courts the most important and fateful of court cases, *Plessy v. Ferguson*. Homer Plessy was a Tremé Creole chosen to test the law segregating rail cars.[20]

Tremé's changing cultural landscape of this period can be seen as having provided Creole (and eventually African American) meaning as refuge from national and racial animosities of Americans and whites, respectively (and interrelatedly), and as a place where Creole society could produce and express its own cultural identities. As Americans became more dominant, the Francophone population turned inward, forming institutions that catered to their social, cultural, and economic interests. The visible presence of these institutions in the landscape speaks to their importance in nineteenth-century Creole society. Beyond the refuge of Tremé's familiar landscape, its institutional concentration, and its often-radical population, the most important New Orleans sociospatial distinction was uptown versus downtown, American versus Creole (French).

This American–Creole New Orleans divide was replicated by the city's African-descended population. In the post-Civil War period, the influx of American blacks or freedmen settled uptown, where they formed their own institutions. The meanings of landscape to each subgroup are apparent in the stories of early jazz musicians. In an interview with Alan Lomax, for example, Creole musician Paul Dominguez reveals, from the Creole of Color perspective, the racial ideas attached to uptown and downtown areas:

> You see, we downtown people, we try to be intelligent. Everybody learn a trade, like my daddy was a cigar maker and so was I … we try to bar jail. … Uptown, cross Canal yonder they used to jail. There's a vast difference here in this town. Uptown folk all ruffians, cut up in the fact and live on the river. All they know—is get out on the levee and truck cotton—be longshoremen, screwmen. And me, I ain't never been on the river a day in my life.[21]

Part II

Although the cultural landscape of Tremé continued to develop during the late nineteenth and early twentieth centuries, the 1920s saw the beginning of another development path for the neighborhood and the city.[22] In this period destruction became the most significant method of landscape change. In the 1920s the city decided to construct an expansive civic and cultural center near Congo Square in the Faubourg Tremé. Tremé had begun to deteriorate in recent decades because of neglect, causing the once mixed neighborhood to experience a white exodus, leaving the neighborhood increasingly black and poor.[23] The civic or cultural center proposal was meant to serve as a catalyst counteracting this neighborhood decline, and although its proposed and actual physical form went through several iterations during the twentieth century, in concept it has remained at the center of redevelopment efforts in Tremé since its initial proposal.

The civic and cultural center, as originally conceived, required demolishing several blocks of homes in both Tremé and the French Quarter. Lack of funding meant that the project proceeded slowly and only in Tremé. The major outcome of the project was the Municipal Auditorium, which sits adjacent to Congo Square. Constructing the auditorium required destroying Globe Hall and a block of historic structures and residences. Located on St. Claude and St. Louis avenues, Globe Hall is recalled by one jazz historian as "the most renowned 19th-century downtown dance hall for colored people."[24] Globe Hall is just one of several significant places lost to the civic and cultural center and subsequent projects.

By the 1930s the French Quarter was also several decades in decline, but it saw a reversal of fortunes with the creation of the Vieux Carré Commission,[25] a regulatory body charged with maintaining the Quarter's architectural character. From that time onward the Quarter experienced increases in property values and the number of commercial enterprises. It also signaled the beginning of the New Orleans preservation ethos, which would have significant impacts on Tremé more than seventy-five years later.

The civic and cultural center complex would be reborn in the years following World War II as the Assembly Center and again in 1968 as a Lincoln Center-style performance complex. By 1970 the various cultural center plans had yielded only the Municipal Auditorium and the Theater for the Performing Arts. Eight blocks of historic housing were also lost as hundreds of families were displaced and significant cultural landmarks including Co-Operators Hall, the San Jancinto club, and the aforementioned Globe Hall were destroyed.

Coincidental to building developments involving the civic and cultural center were plans to route an interstate highway, I-10, down Claiborne Avenue. Claiborne Avenue lacked the built culture of the cleared Tremé blocks, but it served an important function as a tree-lined open public space (neutral ground) and business district. One of the most popular uses of Claiborne Avenue, before the interstate was constructed, was as a community gathering space to experience black Mardi Gras. A 1961 *Times-Picayune* editorial referring to the Claiborne Avenue neutral ground's imminent destruction to build Interstate 10 claimed, "It will take many years to heal the wound represented by the elimination of this grand-style reservation of open-park." The article's enduring sentiment though was that Claiborne's transformation for traffic purposes was "indispensable to general progress."[26]

The final attempt to finish the civic and cultural center complex coincided with the election of Mayor Moon Landrieu in 1970 and the death of Louis Armstrong one year later. A committee appointed by Landrieu to select the site to memorialize Armstrong chose the still largely unoccupied space of the frequently proposed cultural center in the Tremé neighborhood. The idea was to complete the center and to rename it in honor of Armstrong. Proponents deemed the site appropriate because of its proximity to Congo Square and other sites important to the birth of jazz.

The city hired the Baltimore firm Wallace, McHarg, Roberts, and Todd, who partnered with local architect Winston "Robin" Riley to present an acceptable two-phase, revised plan. The first phase consisted of Congo Square renovations, creating a jazz complex, a lagoon and fountain system, landscaping, and parking. During phase 1 construction of the newly named Armstrong Park, the city of New Orleans presented a general management plan to guide urban development. One of the plan's mandates was to create the Historic District Landmarks Commission, an agency charged with overseeing the city's formally recognized historic districts outside of the French Quarter. Although Faubourg Tremé was not a historic district at the time, the commission is implicated in Tremé's present gentrification issues.

Armstrong Park's second phase sought to make the park a contributor to the city's coffers. The city considered two multimillion-dollar plans during the 1980s, both of which proposed amusement park entertainment complexes for Armstrong Park. Tremé's most vocal activists expressed concern that private development might restrict access to the neighborhood's children and the poor. In the end failure to win federal monies cancelled plans for private development of Armstrong Park. The creation of New Orleans's Jazz National Historical Park in 1990 finally provided the city an opportunity to make the park financially and culturally productive. Shortly after the park's creation, its advisory board selected Armstrong Park for the site of its visitor center. Fifteen years after the park's creation, the National Park Service has yet to break ground on its visitor center. Tremé's landscape presently is not undergoing any major changes, construction, or demolition.[27]

Making sense of Tremé's contemporary landscape is a complicated business, based on the neighborhood's current demographic composition. Tremé today is far from homogeneous and so embodies conflicting meanings both internally and externally. With the certainty of oversimplification, it is possible to group these prevailing sentiments. First, the dominant interpretation of the landscape is crafted by the neighborhood's low-income African American residents. They have been the primary residents of the neighborhood since the exodus of whites more than three-quarters of a century ago. Even before the exodus of whites, African-descended people living downtown turned to the neighborhood for their civic and social and spiritual outlets. The most important of these places is Congo Square, where an African presence is linked directly, and arguably, to American popular culture through jazz, and neighborhood bars, clubs, and halls as well as churches and funeral homes have continued their institutional roles in creating meaning for African Americans in Tremé. As continual and long-term residents, people of this neighborhood have repeatedly inscribed the landscape with meaning while simultaneously deriving their identity from it. According to neighborhood resident Norman Smith, "It's all like a web that … had to anchor itself in various aspects in the community."[28] Although linking these various places was accomplished mainly through daily activities, the neighborhood parade also has been important in linking significant places. Parading is a longstanding, annual tradition that although seasonally ephemeral serves to mark space in local memories. Traditionally winding through the streets of New Orleans's African American communities, these parades pick up participants, or second liners, who follow the parade as it stops at important neighborhood places. The tradition, known as second lining, is entrenched alongside other African American traditions, as part of New Orleans's present tourist economy. Since the late 1960s neighborhood activists have fought, often successfully, against perceived intrusions and injustices to Tremé, namely, urban renewal and the interstate highway program, by drawing on the rich political and cultural history of the neighborhood in opposition to citywide racist practices.

The second group is composed of upper-middle-class genitrifiers who are newer to the neighborhood. This group is predominantly but not exclusively white and more well-off than the rest of the neighborhood. This group brings aesthetic concerns to the neighborhood, both in terms of architecture and comportment. Tremé was able to resist the gentrification that occurred in Faubourg Marigny in the 1970s because of a combination of active resistance and the neighborhood's proximity to public housing projects.[29] Presently the neighborhood's proximity to the French Quarter and the central business district and its stock of historically valued architecture, combined with low interest rates, appear to be spurring real estate speculation and gentrification. On occasion those aims and values are at odds with the longer-term residents and traditions. One of the continuing points of contention in the Tremé community is the music played in neighborhood bars. Tremé has a long-standing tradition of live music at its corner bars. Newer residents, upset at the noise level, have sought to quiet these bars by claiming that they are in violation of local zoning ordinances.[30]

The third, and most important, group regarding the future of Tremé may be the combined departments and operations of the City of New Orleans government. The meaning of Tremé for the city is based on the state of the city's contemporary urban economy and postintegration demography. New Orleans's African American cultural traditions, namely, music, food, and street culture, are internationally renowned. These are almost wholly a product of the city's African American culture, and many are directly associated with Tremé. The city is well aware of Tremé's history as it relates to these traditions and thereby has an interest in not turning a blind eye to neighborhood concerns. Furthermore, the community's historical memory forces the city to always face its complicity in the dismantling of Tremé. In theory, that New Orleans is predominately African American, with a black mayor and majority black city council, should also make the community's concerns more resonant.

Recently, considerable efforts have been made by the city to reclaim the formal and informal public spaces of Tremé. After decades of advocating projects that chipped away at both the community and its public spaces, the city has joined with nonprofit and community organizations to promote community building. The city's increased presence in Tremé took place most notably under the mayoral administration of Marc Morial. Morial's director of housing, Vincent Sylvain, targeted Tremé as one of four blighted areas with development potential. Not only was Tremé blighted, it possessed development potential as one of the nation's oldest neighborhoods for people of color; a fact, Sylvain added, that should make black people feel proud, not threatened or embarrassed. The cornerstone of the city's attempt to revive Tremé is the renovated Villa-Milleur mansion on Governor Nicholls Street and the African American Museum that it houses. The city spent $1.2 million to renovate the building and grounds, including the slave quarters. In the years following the city also purchased several adjacent residential properties, converting them into a gallery hall and exhibit space. The museum is meant to "benefit and recognize the Tremé community and African-Americans in particular and their contributions to New Orleans both from a cultural standpoint and an architectural standpoint." The city's effort to improve Tremé's housing included a plan to benefit both the owner-occupied home and the renter. A stated aim was to stem the tide of gentrification and keep traditional residents.[31]

Conclusion

The origins of New Orleans's and Tremé's Afro-Creole landscape can be traced to the eighteenth century, and many authors or agents have shaped the landscape through multiple processes. Regarding those parts of Tremé constructed in the eighteenth and nineteenth centuries, we might ask if the authors were those who constructed the material landscape or those who made the landscape necessary through their racial policies. Similarly, how do we reconcile the meaning inscribed in the landscape at its construction with its contemporary use and meaning?

The mix of residents and meanings in Tremé both naturalizes and challenges prevailing operations of race. From the gentrifiers and preservationists who seek to maximize profit, location, and aesthetics we see the normalization of the discourse that black experiences and histories as embodied in the landscape are expendable. This is an attitude not restricted to whites. As mentioned earlier, the city of New Orleans was actively involved in trying to create a black space in the neighborhood. On the other hand, there are those who seek to "Save Black Tremé." In attempting to preserve cultural traditions and housing options of local black residents, black activists are strategically reinforcing the prevailing racial binary as they try to counter more negative effects of redevelopment, displacement, and devalorization of traditionally Creole and African American sections of the city.

Coda

In August 2005 Hurricane Katrina made landfall in southeast Louisiana as a Category 4 storm. Katrina was not the perfect storm predicted to destroy New Orleans. The outcome was the same, though. Katrina's storm surge filled Lake Pontchartrain and three of New Orleans's canals: two are normally drainage canals and the other is for shipping. Levees on each canal broke. By the time the lake and city water levels equalized, 80 percent of New Orleans was inundated. The greatest tragedy was the slow federal response to the tens of thousands of people trapped in the flooded city with no food, water, or power. At the time of this writing, water is still being pumped out of the city and the dead are not yet accounted for. More than half a million people are displaced around the country, many thousands in shelters in strange cities.

One of the most pressing questions for New Orleans is whether and how to rebuild. Predictions are that after being under polluted waters for weeks, large parts of the city will need to be demolished. With the city's situation vis-à-vis the natural environment not improving, there are scholars and scientists who advocate not rebuilding the city or at least building one substantially smaller and different. If this happens there are fears that many of the city's poor will have no place to return to as developers and property holders seek to rebuild for the middle class.

Faubourg Tremé is presently underwater. Photos taken from atop Interstate 10 looking into the neighborhood showed water at least waist deep. Under those circumstances the questions asked in this chapter, particularly the contemporary questions of saving black Tremé, may seem inappropriate. Like other low-income African American neighborhoods, the location and safety of its residents are unknown. However, when the waters in Tremé recede, the neighborhood's future is at stake. A largely renovated Tremé may no longer be affordable to poorer residents; the same is true for a demolished and rebuilt Tremé. If this becomes the case, the landscape meanings established by the local African American population (and other groups) that are central to questions of personal and neighborhood identity may be lost, as new landscape meanings and interpretations supplant the old.[32]

Notes

1. Essential to understanding discussions of New Orleans is the geographic orientation of the city. In New Orleans, compass directions don't match corresponding sections of the city. For example, the section of city called uptown is not north; generally uptown is upriver and downtown is downriver. The other major directions are not east and west but river and lake. Figure 2.1 presents a schematic and broad overview of the key districts and boundaries discussed in this chapter. This map is unavoidably problematic, as boundaries, districts, roads, and other spatial features of the New Orleans landscape have changed over the two-hundred-year-plus time span that is this chapter's subject (and they continue to do so). Readers familiar with New Orleans are asked to indulge the historical–geographical challenge of representing the particulars of a dynamic city to make sense of the chapter's more general narrative. Thanks to Dick Gilbreath of the Gyula Pauer Cartographic Information Lab at the University of Kentucky for preparing this map.
2. Actually, many houses appear small because of their narrow street frontage. These houses often are situated on long lots that conceal the actual house size from the street.
3. "Back o' town" is another New Orleans directional oddity referring to the parts of town away from the river, which was front o' town.
4. Jerah Johnson, "Colonial New Orleans," in *Creole New Orleans,* ed. Arnold Hirsch and Joseph Logsdon (Baton Rouge: Louisiana State University Press, 1992).

5. Joan Martin, "Placage and the Louisiana *Gens de Couleur Libre,*" in *Creole: The History and Legacy of Louisiana's Free People of Color,* ed. Sybil Kein (Baton Rouge: Louisiana State University Press, 2000).
6. At this time what is now known as the Quarter, or the French Quarter, and the city of New Orleans were the same; the area outside the ramparts were either plantations or not yet developed under European colonial regimes.
7. Haiti, Cuba.
8. Caryn Bell, *Revolution, Romanticism, and the Afro-Creole Protest Tradition in Louisiana, 1718–1868* (Baton Rouge: Louisiana State University Press, 1997).
9. Faubourg St. Marie (St. Mary).
10. Mary Christovich and Roulac Toledano, *New Orleans Architecture: Volume VI; Faubourg Tremé and Bayou Road* (Gretna: Pelican Publishing, 1980).
11. Joseph Tregle, "Creoles and Americans," in *Creole New Orleans,* ed. Arnold Hirsch and Joseph Logsdon (Baton Rouge: Louisiana State University Press, 1992).
12. Richard Campanella, *Time and Place in New Orleans: Past Geographies in the Present Day* (Gretna: Pelican Publishing, 2002).
13. Bell, *Revolution, Romanticism.*
14. Jerah Johnson, "New Orleans Congo Square: An Urban Setting for Early Afro-American Culture Formation," *Louisiana History* XXXII (1991): 117–57.
15. Tregle, "Creoles and Americans."
16. Ibid.
17. Bell, *Revolution, Romanticism;* John Alberts, "Origins of Black Catholic Parishes in the Archdiocese of New Orleans" (dissertation, Louisiana State University, 1998).
18. John Blassingame, *Black New Orleans 1869–1880* (Chicago: University of Chicago Press, 1973).
19. Christovich and Toledano, *New Orleans Architecture.*
20. Rodolphe Desdunes, *Our People and Our History* (Baton Rouge: Louisiana State University Press, 2001).
21. Alan Lomax, *Mister Jelly Roll: The Fortunes of Jelly Roll Morton, New Orleans Creole and "Inventor of Jazz"* (New York: Duell, Sloan and Pearce, 1950).
22. City of New Orleans, Planning and Zoning Commission, *City Plan Report: Civic Art* (chapter IV) and attached letters, prepared by Harland Bartholomew and Associates, 1931.
23. Michael Crutcher, *Tremé Daze* (Baton Rouge: Louisiana State University Press, forthcoming).
24. Richard Collins, *New Orleans Jazz: A Revised History* (New York: Vantage Press, 1996).
25. H. Irvin, "Vieux Carré Commission History: The First Fifty Years," 1998, http://www.new-orleans.la.us/cnoweb/vcc/index.html (accessed April 29, 2002).
26. *Times-Picayune,* "51 of 253 Oaks Will Be 'Saved,'" November 15, 1961. Editorial.

27. Michael Crutcher, "Protecting 'Place' in African American Neighborhoods: Urban Public Space, Privatization and Protest in Louis Armstrong Park and the Tremé, New Orleans" (dissertation, Louisiana State University, 2001).
28. Crutcher, *Tremé Daze.*
29. Larry Knopp, "Some Theoretical Implications of Gay Involvement in an Urban Land Market," *Political Geography Quarterly* 9 (1990): 337–52.
30. Scott Aiges, "Dispute over Live Music Pits Bar Owners and Residents," *Times-Picayune,* March 24, 1993, Metro B3.
31. Crutcher, "Protecting 'Place' in African American Neighborhoods."
32. Michael Crutcher, "Build Diversity," *New York Times,* September 10, 2005.

CHAPTER 3

The White-Pillared Past: Landscapes of Memory and Race in the American South

STEVEN HOELSCHER

Once the owners gave identity to the houses in which they lived. Today, the houses identify their historic shadows. There was majesty here.

—Hodding Carter[1]

Come, take a trip with us across the threshold of time. Leave behind the cares of this moment to stand refreshed again in the long shadows of the great White-Pillared Past that quietly lingers here.

—Reid Smith and John Owens[2]

Natchez: Humid beautiful city, theme park of slavery and the old ways.

—Anthony Walton[3]

First-time visitors to Natchez, Mississippi, might be forgiven for not immediately recognizing the city's majesty. Arriving by historic steamboat plying up the Mississippi or by automobile cruising down the Natchez Trace

Parkway, tourists are often struck by how diminutive and ordinary the place initially seems. At just more than eighteen thousand residents, Natchez barely qualifies as an urbanized settlement and in many respects is a typical New South town of strip malls, modest industry, and quiet neighborhoods. First impressions frequently register on the Isle of Capri Casino, permanently docked on the river shore; the substantial, if matter-of-fact, suspension bridge connecting Louisiana and Mississippi; or the Natchez Finest Plantation Mobile Home Park just off Highway 61 (see Figure 3.1). If a visitor stops at Bluff Park overlooking the Mississippi River, he or she will likely see a historic marker explaining that the town's origins date to French settlement in the early eighteenth century and that, with trade and cotton, Natchez became the "commercial and cultural capital of the Old South" (see Figure 3.2).[4]

If the ordinary landscape of introductory Natchez fails to live up to expectations, it doesn't take long to discover tangible evidence of its Old South past and to see what led the National Trust for Historic Preservation to deem it, in 2003, "one of America's most distinctive destinations."[5] Beyond the kudzu-covered river bluffs, past the African American neighborhoods that are home to more than half the city's population, and amidst the lush vegetation of the Lower Mississippi Valley stands an exceptional collection of antebellum mansions. Guidebooks with color-coded symbols point the way past the mundane to the extraordinary, to the magnificent suburban estates embellished with fanciful names like Dunleith, D'Evereux, Melrose, Elms Court, and Rosalie. White-columned porticos provide the point of entry into nearly three dozen historic homes, all of which are listed on the National Register of Historic Places or are valued as National Historic Landmarks. Distinctive, grand, and opulent, the city's baronial estates are living embodiments of the Old South (see Figure 3.3).

When most people think of the American South before the Civil War, landscapes such as Natchez's storied mansions frequently come to mind: colossal estates built in Greek Revival style, a "white-pillared past" surrounded by fragrant magnolias and luminous azaleas.[6] It is a historic landscape that extends far beyond Natchez, encompassing a broad arc from Virginia and coastal Georgia, extending to Middle Tennessee and eventually stretching to East Texas. Nowhere, however, is the white-pillared past made so explicit on the landscape, and implicit through stories about that landscape, as in the Deep South, especially in that region's historic commercial and cultural capital.

Leading citizens of Natchez work hard at cultivating this image and this landscape, and both embody a powerful invented tradition that says so much about race in America. Natchez and its historic home tours

Figure 3.1 Ordinary landscapes throughout the South, such as Natchez's Finest Plantation Mobile Home Park, frequently invoke the antebellum past, if only in name.

Figure 3.2 Once the "commercial and cultural capital of the Old South," Natchez, Mississippi, today is in many respects a typical New South town of modest industry, quiet neighborhoods, and small public places—like Bluff Park shown here.

Figure 3.3 Originally built during the 1820s and opened for tourists one hundred years later, Rosalie embodies the romantic image of the Old South. The immaculately preserved mansion, together with the more than one hundred other properties listed on the National Register of Historic Places, forms the landscape centerpiece of the region's tourist economy.

crystallize a version of the past immediately recognizable as "the Old South." Ever since 1932, when elite white women of the Natchez Garden Club invited the world to tour their antebellum mansions, the annual Natchez Pilgrimage has become a staple among tourists looking for the place "Where the Old South Still Lives."[7] Historic preservation and economic development by means of heritage tourism have a long, seventy-five-year history here, lending credibility to the claim that a visit to Natchez is, indeed, a step into the past.

A question that immediately arises, for some visitors, is, Whose past are we talking about? Despite the many years of associating the grand antebellum mansion with the golden age of Life before the War of Northern Aggression, recent voices have questioned a singular understanding of these iconic landscapes. One such voice is *A Southern Road to Freedom,* a gospel performance presented annually for the past dozen years as an alternative to the Pilgrimage's primary evening entertainment, the Confederate Pageant. *Southern Road*'s narrator asks the audience to "contemplate the many antebellum mansions that [they] will see on tour" and to remember that "Stanton Hall was built by free black and slave labor." He goes on to remind the audience,

> The impressive white columns of Dunleith are not just symbolic of white supremacy and oppression, but also of the hard work by African Americans. Not only did we build it, but also one of Mississippi's most important politicians—John Lynch, who, at the age of 25, rose to become the youngest man to serve in the U.S. House of Representatives—once fanned dinner guests as a slave

> in Dunleith's dining room. The many antebellum mansions speak to the hard work of free black and slave labor.[8]

Symbols of opulence and gracious living for some become markers of subjugation for others. Landscapes of "romance, grandeur, chivalry, and wealth" are revisited as hallmarks of pain and coerced labor as Natchez's white-pillared past is increasingly seen through the prism of race.

More generally and in a continuous expanse across the American South today are cars with bumper stickers proclaiming "Heritage, Not Hate" that sit uncomfortably next to those carrying protestors with signs declaring "Your Heritage Is My Slavery." In courtrooms and in shopping malls, Civil War reenactors clash with civil rights advocates over renaming downtown streets after Martin Luther King Jr. And state governments in South Carolina, Georgia, Mississippi, and Texas become embroiled in controversies over public landscape displays of the Confederate flag. In Natchez and throughout the American South, the past and present tangle in a complex web of political, economic, and cultural relationships that speak to the region's ongoing struggle of identity. Landscapes of race and memory stand at the symbolic and political–economic center of that struggle.[9]

Natchez is an excellent place to examine those landscapes not only because of the Pilgrimage's contemporary importance—it's a huge moneymaker that annually draws more than a hundred thousand tourists—but also because the small Mississippi city became the first to stage an event that has since become a significant regionwide movement. Tourists from around the region, country, and world flock to places like Natchez to tour its plantation home-museums, infusing significant amounts of capital and social legitimacy along the way. Unlike the New South after Reconstruction, however, today's newest, suburban South hears challenges to a perspective that, until recently, few felt able to oppose directly.

Expressing "the Bone and Sinew of the State"? The Evolution of a Landscape Ideal

The plantation mansion in the American South—no less than the New England village, Main Street of Middle America, or the California suburb—has achieved the status of national symbolic landscape.[10] From Joel Chandler Harris's Uncle Remus stories at the end of the nineteenth century to Tom Wolfe's best-selling novel *A Man in Full* at the end of the twentieth century, the southern manorial estate became a firm fixture of American mass culture. A monumental white house of "perfect symmetry ... tall of columns, wide of verandas, beautiful as a woman is beautiful who is so sure of her charm that she can be generous and gracious to all" is

Figure 3.4 The Hollywood version of Margaret Mitchell's *Gone with the Wind* instantly turned a well-known regional landscape into an internationally recognizable icon. Tara, pictured here in a water color set design by William Cameron Menzies for David O. Selznick's blockbuster 1939 film, was more than merely an antebellum house: it is emblematic of a national symbolic landscape. Reprinted with permission by the David O. Selznick Collection, Harry Ransom Humanities Research Center, The University of Texas at Austin.

how Margaret Mitchell famously described this landscape. Such imagery achieved its most memorable expression with the spectacular 1939 Hollywood version of Mitchell's *Gone with the Wind,* arguably one of the most enduring and popular movies ever produced (see Figure 3.4).[11]

A white-pillared past is as evident on the land as it is in popular books and on the movie screen. Even the most mundane landscapes incorporate elements of manorial ideal: welcome centers greeting visitors at state lines throughout the region consistently deploy a quasi-Greek Revival building style; new subdivisions in old southern cities frequently bear the name "Plantation Trace," "Windsor Hill Plantation," or "Bayou Creek Plantation"; and businesses, large and small, across the region are connected, if only in name, to their antebellum predecessors (see Figure 3.1). Especially as embodied in the Ur-American spaces of Mount Vernon and Monticello, the southern plantation mansion can rightly be counted as "part of the iconography of nationhood, part of the shared set of ideas and memories and feelings which bind a people together."[12]

The shared set of ideas, memories, and feelings evoked by this landscape arose not from the rich loess soil of the Lower Mississippi Valley but from the equally fertile pens of writers, politicians, and community leaders committed to constructing an imaginative geography of the region's past. The southern plantation ideal is woven so deeply into the fabric of the

American symbolic landscape that otherwise sure-footed scholars have occasionally abetted its popular image. The architectural historian Alan Gowans, for one, maintained that "for the southern aristocrat, Greek and Roman architecture was the symbol and the assurance that a sound society could perfectly well combine the ideals of liberty with the institution of slavery. … Greek temples were in southern eyes practically a statement of the southern way of life."[13] Neoclassical architecture, so the argument ran, was a natural embodiment of the plantation owners' desire to reconcile American democracy with chattel slavery.

As a tangible landscape, however—a "concrete, three-dimensional shared reality," as J.B. Jackson would put it—the spectacular Greek Revival mansion was a distinctly uncommon sight in the antebellum South.[14] John Michael Vlach reported that only the plantations run with very large numbers of slaves (that is, a hundred or more) approximated the manorial ideal. With that in mind, Vlach estimated that a little more than a thousand plantations throughout the region "developed to the state of elegance promoted by the widespread southern mythology." What is indeed remarkable is how "such an unrepresentative place as the great plantation estate came to dominate the self-perception of the South."[15]

Identifying the region with this landscape came only with time—and a great deal of work. Before the Civil War, wealthy southern planters, to be sure, enthusiastically built homes following neoclassicist conventions, but such stylistic conventions were one set of many trends in a diverse region. The South was home to a variety of peoples and cultural landscapes: although slavery might have been the economic centerpiece of the region, even the peculiar institution could no more impose a uniform way of building than industrial capitalism could to the north.[16] Classicism, as the first national style of the United States, was far from an exclusively *southern* building tradition, as many dwellings and public structures across the trans-Appalachian frontier—from Illinois and Michigan to Alabama and Mississippi—reflect this tradition.[17] And by the 1850s—the decade of the most strident calls of southern nationalism—Greek Revival was on the decline in the South as more eclectic styles came into fashion, styles that came under the name of Italian villa, neo-Gothic, and Moorish.[18]

Indeed, among defenders of southern nationalism during the crucial decade before the Civil War, the Greek Revival mansion seemed out of step with the cause of southern geopolitical identity. One writer for the patriotic agricultural journal *Southern Cultivator* found that "Grecian architecture is decidedly unfit for domestic purposes. Its staid pillars, unbroken shadows, its heavy frowning entablature looks uncongenial."[19] Another promoter of southern nationhood condemned the extravagance of erecting "what would appear to be massive columns, but which are generally made of wood, in the ridiculous ambition of appearing to live in something of a Greek temple." Such a structure might be appropriate for public building, but, he added, for a dwelling "a worse taste can hardly be imagined."[20] Most agreed with Charles A. Peabody, who advised southerners to build simple houses adapted to the environmental conditions of the South, to build "not a gew gaw palace or ginger bread cottage, but a substantial, comfortable house." Landscape might be adopted to express "the bone and sinew of the State," but it surely would *not* be found in the "gentlemen's country residences and town mansions."[21] Visitors and local residents may have viewed the Greek Revival mansion with a mixture of admiration, envy, and deference; certainly, the giant wooden or stucco-covered brick colonnade featured at the entry to the manorial estate was a bold announcement of status and success. But little evidence suggests that southerners before the Civil War saw such landscapes as a symbol for any southern way of life.[22]

Inventing a Southern Tradition: the Old South Framed by White Pillars

This was to change during the post-Reconstruction era known as Jim Crow, when southern elites invented a powerful tradition of the plantation mansion landscape.[23] That tradition, which I am calling the white-pillared past, crystallized at a moment of rapid social and political–economic change. For its social legitimacy and to give it the weight of material reality, that tradition was founded on the highly charged landscapes of race.

By the 1880s political conflicts over the meaning of freedom for blacks as well as economic–geographic trends toward centralization, standardization, urbanization, and mechanization meant that American collective identity was anything but clear. White southerners, like Americans in other regions, sought to make sense of the apparent fragmentation of their world by envisioning new foundations for identity. Such foundations for white southerners seeking social order were invariably spatial, with the imagined historical geography of the region's landscape offering the prime source material; they were also inevitably political, with the fiction of absolute racial difference becoming the region's governing principle.[24]

At bottom, those foundations—or imaginative geographies—were not constructed wholly out of new cloth but constructed from the hazy boundary separating experience from myth. Nearly a half century ago, C. Vann Woodward put it cogently: "The twilight zone that lies between living memory and written history is one of the favorite breeding places of mythology. This particular twilight zone [of the southern landscape] has been especially prolific in the breeding of legend."[25] More than just a reaction to a turbulent world where Civil War defeat destabilized categories of power and authority, the white-pillared past became an active ingredient in defining life in the New South; it became as much a source of division and conflict as "a shared set [of] ideas, memories, and feelings." For many Americans—northerners and southerners, blacks and whites—Natchez became its culture hearth.[26]

The landscapes of memory in Natchez, as in other southern cities, followed a distinct, if at times overlapping, periodicity. First, immediately following the Civil War, commemorations of "the Cause that could never be lost" built the Confederate monuments that today can be found at the center of countless southern towns and cities. With its emphasis on crafting its own history of the "war between the states," the Confederate or Lost Cause phase of southern remembrance may have dominated regional memory shortly after Appomattox, but by 1920 the movement had lost much of its influence.[27]

Beginning in the 1880s and gathering momentum during the decades surrounding the Great Depression, a second phase of southern memory pushed time backward to the period before the War of Northern Aggression, to the Old South. "Perpetually suspended in the great haze of memory," W.J. Cash wrote, the image of the Old South "hung, as it were, poised, somewhere between earth and sky, colossal, shining, and incomparably lovely—a Cloud-Cuckoo-Land."[28] Cash and other debunkers of the Old South tradition identify several elements that embody this cultural memory: the essentially precapitalistic and feudal characteristics of the plantation; the hierarchical ordering of plantation relationships, descending from the wellborn father to the lowliest slave; the manners, behavior, and demeanor of all society's members that emphasizes joviality and conviviality; and, most important of all, the idealized relations between the races.[29]

The towering columns of the neoclassical plantation mansion framed this picture of the Old South and gave it substance. Writers such as Joel Chandler Harris and Thomas Nelson Page, publishing best-selling novels and short stories in such popular magazines as *Harper's, The Century,* and *Scribner's,* were instrumental in creating a widely understood vision of the Old South. For Harris, the image of the southern plantation was unambiguous. At its center stood

> a stately house on a wooded hill, the huge white pillars that supported the porch rising high enough to catch the reflection of a rosy sunset, the porch itself and the beautiful lawn in front filled with a happy crowd of lovely women and gallant men, young and old, the wide avenue lined with carriages, and the whole place lit up, as it were, and alive with gay commotion of a festive occasion.[30]

Page expanded on this idealized landscape, casting it in a mellow light and reflecting on the social relations that supported it. He may not have produced distinct Figures like Uncle Remus and Brer Rabbit, and he certainly possessed none of Harris's "secret racial subversiveness,"[31] but Page's nostalgic stories, novels, and poems of the pre-Civil War South, as well as his historical and political essays, remain the standard of plantation literature genre, making him an even more influential architect of the white-pillared past. "Life on the [plantation] was amazing," Page insisted in his aptly titled book *The Old South*. "There were the busy children playing in troops, the boys mix[ing] with the little darkies as freely as any other animals, and forming associations which tempered slavery and made the relation one of friendship."[32] Such essays invoked intense nostalgia for the plantation era and hint at Page's unmatched ability to glorify landscape and social life before the war. They also signal his greatest contribution to constructing the Old South: the enduring fiction that steadfast kindliness characterized race relations before the war.

"Columns and Live Oaks and Mammies": Natchez's Landscapes of Memory

It was one thing to read about the white-pillared past in the American South, when blacks purportedly knew their place of subordination within a benevolent social order and where the supremacy of white power remained unchallenged. It was quite another to see the landscapes epitomizing this imaginative geography. Here is where Natchez and its twice-yearly Pilgrimage become central. As the sites through which stories and rituals of citizenship are enacted and resisted, landscapes are especially powerful media for dominant groups to present their case to the world. To be believable, inevitably selective versions of the past are often made concrete through material objects. The ephemeral nature of memory requires their architects and guardians to ground memory in physical form—in landscapes. A landscape, Fred Inglis noted, provides "the most solid appearance in which a history can declare itself."[33]

Pierre Nora extended this argument by contending that memory "relies on the materiality of the trace, the immediacy of the recording, the visibility of the image." Landscapes of memory—the monuments, memorials, and museums that anchor collective remembrances and make them user-friendly—are the sites where, as Pierre Nora put it, "memory crystallizes and secretes itself."[34] Cultural landscapes, through their compelling symbolic qualities, help make a social order appear natural and even, as Richard Schein argues, "normative," or a reflection of how the world *ought* to be.[35] Landscapes contrasting the splendor of the antebellum mansion with the squalor of poor black housing, as guidebooks to the city often did, seem to offer irrefutable proof of what the *Natchez Democrat* could only assert; namely, that "history shows conclusively that certain individuals and certain races are superior to others."[36]

The elite white women of Natchez's Garden Club understood the symbolic power of landscape better than most, and certainly more than their businessmen husbands. In response to the economic pressures of declining cotton production, Natchez's white businessmen pursued a path of modest industrialization, emulating Henry W. Grady, the influential editor of the *Atlanta Constitution,* and his vision of an industrial New South divorced from its antebellum past. Natchez's elite white women mined a different resource, one that, over the years, has proved much more profitable. One of the Garden Club's leaders put it this way:

> Have you ever listened to sentimental songs and tales of the Old South that made you long for a glimpse of white-columned plantation houses and moss-draped live oaks and moonlight and mocking-birds and faithful old mammies? Have you wondered, and perhaps inquired, if somewhere in the South you could still find a remnant of those mellow and romantic days of "before the war"? Most people will tell you no; the Old South is gone, crushed by the pressure of the machine age.

Roane Byrnes went on to reassure visitors to Natchez that it's not too late:

> For if you will follow the course of the Father of Waters far down into southern Mississippi, you will come ... to one last remaining foothold of the vanishing order. Set high on wooded bluffs overlooking the broad Mississippi River and the far-reaching, fertile lowlands of Louisiana, is the old and picturesque town of Natchez. Here you can still find the Dixie of song and storybook—columns and live oaks and mammies and all.[37]

By marrying tangible, material elements of the antebellum past to stories of its greatness, Byrnes and the other leaders of the Natchez Garden Club successfully turned their city into one of the South's most enduring tourist attractions (see Figure 3.5). How they accomplished such a striking success, especially in the depth of the Great Depression, is a story of tenacity and organizational savoir faire. It also demonstrates the attraction that this landscape of race holds for the scores of sightseers who visit the city annually.

Thanks, in part, to early capitulation during the Civil War, Natchez is home to more grand antebellum homes than any other U.S. city. Its early founding, staggering planter wealth, and unique urban concentration

Figure 3.5 Beginning in the early 1930s, owners of Natchez's antebellum mansions have opened their homes to tourists from around the world. Mrs. S.B. Laub, a member of the Pilgrimage Garden Club, was one such woman who helped invent the Natchez Pilgrimage tradition. This 1937 photograph, taken by David O. Selznick's *Gone with the Wind* research staff, shows Mrs. Laub in front of her home, "the Burn." Reprinted with permission by the David O. Selznick Collection, Harry Ransom Humanities Research Center, The University of Texas at Austin.

combined with early twentieth-century economic stagnation to preserve the structures that in other cities were destroyed. "Natchez, with its outstanding architecture," the National Park Service declared, "stands alongside Savannah and Charleston as a city known for its preservation successes."[38]

Elite white women—not nameless social forces or timeless tradition—were the active agents in this preservation success that has become the economic linchpin of a tourism industry annually drawing more than 230,000 visitors and generating an estimated $83 million. With more than 40 plantation homes and antebellum mansions, and 140 properties listed on the National Register of Historic Places, heritage tourism in Natchez has expanded well beyond the two relatively short Pilgrimage seasons to become the year-round staple of the economy. Although denied formal avenues to power, these women, like their equivalents in Charleston, Nashville, and San Antonio, played a fundamental role in shaping the memory that would instruct and bolster southern politics and economy during the middle decades of the twentieth century.[39]

The principal "guardians of tradition" grew out of the Women's Club of Natchez and called themselves the Natchez Garden Club. Their 1927 charter spelled out the new civic improvement club's three-part mission:

> To promote and foster the beautification of the City of Natchez, its houses, gardens, public buildings. ... To foster and promote a love of the beautiful in architecture, interior decorating and landscaping. [And] to perpetuate the history of the Natchez Territory and to keep alive the memory of the lives, traditions and accomplishments of the people who made that history.[40]

Beginning in the early 1930s and continuing today, the Natchez Garden Club and its competitor, the Pilgrimage Garden Club, have been extremely influential in restoring the city's antebellum homes and in making these landscapes "must see" tourist destinations.[41] When early Pilgrimage leader Katherine Miller took her illustrated lecture on the road, she conjured images of a lost Eden where economics took second place to matters of culture and refinement. Dressed in an antebellum-era hoopskirt, she usually began by entreating her audiences to "take yourself back from these days of modern homes and gardens and go with us in imagination and memory to the glory and grandeur of the Old South as we re-create for you the atmosphere of the antebellum mansions and the beautiful natural gardens of Old Natchez."[42]

Thanks to such tireless promotion by the garden clubs, as well as an avalanche of free publicity with both the novel and celluloid version of *Gone with the Wind*, by 1939 the small city had become, as *House and Garden* put it, "the 'Mecca of charm' for the nation." The interaction between Natchez and *Gone with the Wind* was far from one-way, however, and suggestive of an intertextual relationship between the Mississippi town and Hollywood's blockbuster film. Research for the movie took the producer and his research staff to plantations across the South in search of the ideal landscape. Eventually, they arrived in Natchez, where Pilgrimage hostesses opened their grand mansions to Selznick's researchers. There, they found what they were looking for: a perfect model for *Gone with the Wind*'s most potent symbol, "a graceful, white columned, romantic house"[43] (see Figure 3.5).

Touring Landscapes and Telling Stories of the White-Pillared Past

A fundamental component of Natchez's success as an articulation of the white-pillared past, and as an increasingly important national tourist attraction, were the stories told about these landscapes. The soaring white columns of Natchez's antebellum mansions, though impressive to the eye, remain themselves mute to whatever messages their sponsors hoped to convey. A landscape's meaning does not come neatly packaged, inherently ready to be deciphered or to be simply "read" as a transparent and unproblematic text. Stories need to be told and linked to the landscape so that the landscape embodies the stories.[44] More extraordinary than the architecture of the "old homes," to the journalist John Gunther and to generations of tourists to Natchez, were the actions, gestures, and words of the home owners, which seemed to condense "in their extreme form the basic issues of the white–black conflict."[45]

The stories of the Old South in Natchez are told most compellingly during the biannual Pilgrimage, when the most prized and restored antebellum mansions are opened for touring. Exactly what qualifies a home for inclusion on tour—as an "old home"—is unclear. Certainly the age of the original dwelling and a tangible link to the time before the war are vital. Simple possession of an antebellum dwelling is not enough, however. The most important qualification seems to be membership in one of the two garden clubs, both of which maintain exclusive control over the heritage display. If there was any question about this fundamental structure, it was laid to rest by the recent refusal of Natchez Pilgrimage Tours (the incorporated entity that works on behalf of the garden clubs to manage the event) to include, as part of the Pilgrimage, a nineteenth-century home owned by an African American woman.[46] Even the William Johnson house, once owned by the Natchez Garden Club before the National Park Service acquired it, has been distanced from any connection to the Pilgrimage. Because of its association with Johnson, whose diary represents the most

Figure 3.6 Miss Alma Carpenter greets a group of tourists to her home, "The Elms," during the spring 2000 Pilgrimage. "My family has lived here since 1878," Miss Alma informs her guests, "so I guess that's long enough to call it home."

complete account of the life of a free African American in the pre-Civil War South, the house is one of the most historically significant in Natchez.[47] Exclusion of historic structures such as the Johnson home or, say, the boyhood home of the novelist Richard Wright follows an internal logic: only dwellings that conform to the dictates of the white-pillared past have a place in the Natchez's landscapes of memory; all others are deemed "out of place" and a potential threat to the entire project.[48]

Stories linking the Pilgrimage tour dwellings with the Old South rely on a narrative of authenticity based on the connections that can be established between patrician families and the houses. Most of the tour houses are no longer owned by families of the "old aristocracy," those upper-class families who, as a group, felt that they had "a certain claim upon all of the 'old homes.'"[49] But in the case of houses like Landsdowne, the Elms, or Green Leaves—homes that are occupied by families dating to the antebellum years—tour guides and home owners alike are quick to establish such genealogical connections (see Figure 3.6). And when an "old home" like The Briars is fortunate to count the woman who eventually married Jefferson Davis as one of its earliest inhabitants, lineage trumps all other stories.[50]

Closely related to narratives of patrician families are those that celebrate the women of the Old South home. Inevitably depicted as "a great beauty" cultivated in the finer skills of entertaining European guests and choosing the right color scheme for each room, the Natchez lady was also said to possess great fortitude in the face of adversity. Often she is said to single-handedly save the home from marauding Yankees during the War between the States. At Rosalie, for example, tourists hear that the lady of the house was so unfailing in her support of the Confederacy during the occupation of Natchez that she was forcefully banished to Atlanta. Her contemporary at Montaigne was not so lucky, as "newly freed slaves and white scalawags" together nearly destroyed the home and much of its "beautiful furnishings."[51]

Fortunately, a "number of handsome pieces were saved" because at Montaigne, and at every museum-house, antique furniture reigns supreme. No museum-home tour is complete without a detailed presentation about dozens of unique, historical items. Every Baltimore desk, Philadelphia Pembrooke table, and Sheraton sofa is one of a kind; every Waterford chandelier, set of Old Paris china, and Empire bookcase is priceless. Guides become extremely knowledgeable about styles, periods, and functions of nineteenth-century furniture, and when a difficult question arises from a tourist, less-experienced docents seek out those with more expertise for an answer.[52]

Such stories of the white-pillared past—narratives that focus exclusively on a patrician lineage, specific gender roles, furniture, and architecture—invariably come at the expense of the people who made such extravagant landscapes possible. Because each museum-home was created and sustained through the institution of slavery, the near complete absence of a black perspective during the "old home" tours is astonishing.[53] When Old South blacks are mentioned on tour, they are practically always called "servants," not slaves, and are described in context to the objects they handled: the cook would put this Dutch oven in the fireplace over there, or a servant boy would sit here and pull the shoofly fan over the table. During a tour of Monmouth, for example, it was only when one of my students asked about slaves did we discover that a $200 per night luxury bed-and-breakfast lodging off the main building once housed "servants" before the War (as in the Civil War).[54] And at no time during a tour of Longwood—a one-of-a-kind octagonal house that was never completed—does one hear that its owner, Haller Nutt, achieved the wealth to build his dream home with the labor of his more than eight hundred slaves who lived on twenty-one separate estates from Natchez to the Louisiana Gulf Coast.[55] Touring the grand antebellum homes that give shape and substance to this most traditional outpost of the Old South today would lead one to suspect that enslaved African Americans were not part of this landscape's history.

Challenging Old South Landscapes: Contemporary Black Counternarratives

This has not always been the case. During the Pilgrimage's first three decades, from the early 1930s until the 1960s, African Americans were visibly present in the landscape, on mansion tours, and as performers during the Confederate Pageant. One early description suggests what tourists might expect to find during a plantation home visit: "Perhaps a grizzled, bent, old ex-slave stands to bow you in, or a strapping, courteous young Negro will direct the parking of your car. At another place as you step up onto the gallery, a little colored boy stoops and wipes your shoes. … Awaiting inside to receive you with gracious courtesy, stands the hostess with a group of her friends."[56] Of utmost importance was the juxtaposition—and implicit hierarchy—in the Natchez landscape between white and black, between master and servant, between a storytelling, gracious "old home" hostess in "quaintly beautiful clothes" and silent, subservient "mammies" with "white aprons over plaid dresses" and their "pickaninnies dancing and jigging to colorful tunes of Negro bands."[57]

Such crude binaries, as essential as they were to reinforcing the Jim Crow hegemony of white supremacy, proved unsustainable during the tense atmosphere of the civil rights era. Both the Confederate Pageant and the "old home" tours became untenable symbols of oppression during the South's Second Reconstruction and its black political enfranchisement.[58] The tourist attractions achieved notoriety among the local chapter of the National Association for the Advancement of Colored People (NAACP) for what one activist called their "degrading" depiction of African Americans.[59] In the wake of successful, high profile, 1964 efforts to desegregate Natchez's public facilities, the NAACP and the Student Nonviolent Coordinating Committee took aim at the living and symbolic personification of the Old South. Freedom workers found the white-pillared past to be a harder nut to crack, however. Garden club members, aided by lawyers and armed security guards, denied protesters access to the Pilgrimage events, which were effectively deemed to be taking place in "private," not "public," spaces.[60] Although civil rights protestors might have been unsuccessful in dislodging the Pilgrimage from its central position in Natchez's civic life—if anything, heritage tourism based on a spurious vision of the Old South has only increased during the past forty years—they did effect the removal of objectionable scenes from the pageant and museum-home tours that many found so offensive.

Now, after a lapse of nearly three decades, African Americans have returned to Natchez's tourist front stage but under conditions more in step with the tenets of contemporary multiculturalism than the scourge of white supremacy. During the pilgrimage and through their own historic preservation efforts, black Natchezians have begun telling public stories—counternarratives—of the city's landscapes of memory and race. Although it is beyond the scope of this chapter to offer a systematic comparison of these landscapes, I describe two projects currently underway in Natchez that provide a distinct challenge to the white-pillared past.

A Southern Road to Freedom

A Southern Road to Freedom owes its origins to a contemporary public relations problem: the absence of African Americans from the tourist spectacle became conspicuous for a city that claims to offer visitors the opportunity to "enter a time machine to be transported back to the Old South."[61] It is abundantly clear to African Americans, and to a growing number of whites, that the white-pillared past version of Natchez history told on museum-home tours reflects a skewed account of the past. "Visitors would come to Natchez from other parts of the country and they were amazed that African Americans were entirely missing from the event," one person involved with the Pilgrimage told me. "The whole thing was really an embarrassment." Moreover, African American heritage tourism focused on civil rights and antebellum history is a growing sector of an important regional industry developed since the 1980s, and ignoring blacks also hurts the Pilgrimage at the cash register.[62]

It was in this context that several liberal whites began urging the two garden clubs to reconsider the noticeable lack of black community involvement in the Pilgrimage. They contacted an elementary school teacher with well-known interests in local African American history to help broaden the event. Ora Frazier's response was a tentative "yes, but only if we could tell our own story from our own perspective." That meant, in contemporary Natchez, that black involvement had to avoid the Confederate Pageant—a heritage display so charged with racial antagonism that any direct reconciliation seemed unlikely.[63] The result of these negotiations is *A Southern Road to Freedom,* a gospel performance, now in its second decade, that disputes the narratives and landscapes that have long dominated interpretations of the Natchez past. It provides a compelling alternative.

The white Doric columns found at the entrance to many of the city's antebellum mansions also are found supporting the pediment of Natchez's Greek Revival-style First Presbyterian Church, which has been home to many of the region's most prominent and wealthiest planters since 1829. Members of the Holy Family Catholic Church Gospel Choir transform this historic structure for five weeks every year when, for the past dozen years, they have presented an uncompromising tone of defiance against the sanitized history offered elsewhere in Natchez. "The songs, chants, moans, groans, and hymns that [the audience] hears is the music that extends from the captivity of African Americans during the time of slavery to the present day," the evening's narrator informs the crowd of mostly white tourists. These songs, he continues, "are a testament to the ability of blacks in Natchez to create their own culture, to survive oppression."[64]

Alternating first-person narration of historic black Natchezians and key institutions with stirring gospel songs, *A Southern Road* frames its performance with the tropes of dignity, struggle, resilience, and survival. Enacting stories of resistance, of overcoming oppression, and of self-esteem in the midst of subjugation punctures the illusion of the Old South as a time of blissful race relations when everyone knew his or her "place." Nowhere to be heard are the by-now familiar themes of the contented and faithful servant or the moonlight and magnolia bliss of the Days before the War. And the various scenes are not presented in generic fashion, as the "Fox Hunt" or "Farewell Ball" as one sees at the Confederate Pageant. Instead, the performance personalizes the enslaved and oppressed by giving them names—Abrahima, Jane Johnson, Alice Sims, John R. Lynch, Richard Wright—and by dramatizing their individual experiences and feelings. It

Figure 3.7 Once resembling "a sprawling prison camp," Natchez's Forks of the Road marks the location of Mississippi's most active slave market. Now, after more than a century of neglect and forgetfulness, civil rights activists like Ser Seshs Ab Heter-C.M. Boxley are leading a successful campaign to highlight the significance of this landscape of memory and race.

demands that audience members reconsider the landscapes they have just seen and, equally significant, those they have passed by without notice: "All the symbols of Natchez are important: the large houses of free blacks like Robert Smith and William Johnson; the antebellum mansions; the small houses of freedmen; the small slave dwellings on plantations."

Perhaps most radical of all, *A Southern Road* inverts the cultural memory of the Civil War. The war is performed as a key moment in an ongoing struggle for liberty rather than the end point for a charming way of life. Unlike Natchez's white cultural memory, with its tidy separation of past and present, black countermemory interprets history as an ongoing, unresolved process. Enslavement, repression, and positioning one race over another no longer can be understood as "the most obvious thing in the world": *A Southern Road to Freedom* upends the white-pillared past by defamiliarizing it.[65]

Forks of the Road

Less than a mile east of downtown lies an inconspicuous intersection—seemingly one of hundreds, each just like the other, with a muffler repair shop occupying one corner, a parking lot on another, and streams of automobiles passing by as quickly as possible. This is a return to everyday Natchez, within sight of several colossal mansions—"Linden," "D'Evereux," and "Monmouth"—but apart from them in the landscape's utter ordinariness. Arguably, it is also the city's most historically significant slice of real estate, for, without this landscape, none of the others that so beguile tourists would have been built (see Figure 3.7).

"This is sacred ground," said Ser Seshs Ab Heter-C.M. Boxley, "a holy site. Right at this place is where the enslavement trafficking of human beings found its necessary market." All that remains of the most important slave market in Mississippi's most active slave-trading city are the roads that mark its intersection. Between 1823 and 1863, tens of thousands of enslaved men, women, and children passed through that intersection, known then and now as the Forks of the Road. It was a landscape of considerable importance to the city and one that attracted much attention.[66] Extensive holding pens were constructed to detain the hundreds of enslaved people that slave traders regularly assembled here. Support personnel were required to provide surveillance and to oversee the distribution of food, water, and clothing. Because the winter months were the traditional season for slave trading, barracks were necessary for slaves awaiting purchase. The entire scene, wrote two historians, must have resembled "a sprawling prison camp."[67]

During the long years following the Civil War, Forks of the Road was the kind of landscape that societies work hard to forget. It became, as Ken Foote put it, a landscape of obliteration, where groups in power "actively effac[e] all evidence of a tragedy to cover it up or remove it from view."[68] Street names were changed; the cisterns, holding pens, and barracks were destroyed; new structures were added, including a cotton gin, a grocery store, various dwellings, and warehouses; and the roads were regraded to take a slightly different path. At the same time that elite white women memorialized and sanctified the city's antebellum mansions by turning them into sites of pilgrimage, their neighbors obliterated the shameful landscape on which human beings were bought and sold.

For African Americans like Ser Boxley, the political–economic interactions that occurred at the Forks of the Road make it hallowed ground and a place no less powerful than consecrated battlefield sites like Gettysburg or the Little Bighorn. The landscape, he argued, "permeates the presence, spiritual life, arts, history, legacies, heritage, culture, and humanity of … enslaved Africans in America." A native of Natchez who studied urban planning and who worked for thirty-five years in various antipoverty programs in California, Boxley returned to his hometown in 1995 with the intention of leaving his possessions in storage before his planned move to Africa. His retirement plans quickly changed. "I couldn't believe what I saw—here was this town making its living off a one-sided story of mansions and pageants and whatever else would sell to tourists. It angered me, and it still does." As a longtime civil rights activist, Boxley began leading the fight for what he called "history and tourism democracy," a struggle that moves "beyond the Civil Rights Human Rights Fight" to take on the white-pillared past.[69]

That struggle has evolved during the past ten years, from a small, grassroots movement to one that has attracted national attention. Significantly, it has garnered the support of numerous organizations—from Boxley's own Friends-of-the-Forks-of-the-Road Society, the Historic Natchez Foundation, and the City of Natchez to the National Park Service and the State of Mississippi—that have cut across racial lines. In June 1998 and with funds raised by the Natchez Juneteenth Committee, the Mississippi Department of Archives and History erected a state historic marker at the intersection, which has become the focal point for annual Juneteenth celebrations.[70] Then, with a "prodding, pushing, feather-ruffling, and shaking-people-up" form of grassroots organizing, Boxley led the campaign to acquire the property and bring it into the public domain to be preserved and interpreted. That effort received major assistance in the form of a $130,000 grant from the State of Mississippi to purchase the land from the three property owners, who, at first, proved quite resistant to the offer. Complex and, at times, heated negotiations between the city, the property owners, and preservation groups eventually led to the acquisition of a narrow slice of the land at one of the intersection's two historic "Negro Marts."[71]

Very little remains of the antebellum landscape. In historic preservation terms, the site has been "massively disturbed," making it extremely difficult to place the property on the National Register of Historic Places, much less achieve National Historic Landmark status. This is not an inconsequential dilemma, because Landmark status would go a long way toward integrating the site into the Natchez National Historic Park and to creating a center for the study of the interstate slave trade—the ultimate objective for those most involved.[72] But it also points to a larger issue that underscores the ongoing tensions surrounding landscapes of memory and race in the South: just because a landscape has been obliterated does not mean that its memories have been erased. They just might take a little more effort to see. "Historic white landscapes in Natchez can be summarized by what I call bricks and furniture preservation—let's see all the fancy building styles and furniture styles that make a mansion look the way it does," Boxley explained.

> For African Americans, it's completely different. The important thing to know is who made the bricks, who did the work that made it all possible. That is exactly why the Forks of the Road is such a significant place: the spirits of our ancestors reside there. The pain and suffering of generations of enslaved Africans in America haunt that ground.[73]

Beginning with a fundamentally different premise about what constitutes a historic landscape, the Forks of the Road, like *A Southern Road to Freedom,* offers a necessary corrective to the bias and distortion of city's antebellum museum-homes.

Conclusion

Natchez, long the authorized repository for the white-pillared past—the South's official "theme park of slavery and the old ways"—finds itself at a necessary fork in the road. Although black preservation activists might have to wait for the Forks of the Road slavery site to be listed on the National Register, their goal of a more inclusive history inches closer. The recently elected mayor of Natchez, the first African American to hold the office since Reconstruction, has made it a priority for owners of the city's antebellum mansions to "present a balanced picture of plantation life."[74] In early 2005, after a federally funded $2 million restoration and under the auspices of the National Park Service, the William Johnson House opened to the public with a mandate, in part, to interpret the history of slavery. And in December 2004, after nine years of what Boxley called "relentless agitation" and with the assistance of a $16,000 grant from the National Park Service, the activists succeeded in erecting an extensive, outdoor exhibit of photographs and text dedicated to the memory of African American enslavement.[75]

"It's about time Natchez and other cities in the South began interpreting the historical landscape from more than just one perspective," reflected Jim Barnett, the director of historic properties at the Mississippi Department of Archives and History. "The city has been justifiably criticized for telling a one-sided history, but by recognizing and interpreting this site, that will be harder to do." Certainly, much is at stake in the South's recent battles over the past. If Natchez is to expand its tourism–economy base, it seems essential to broaden its appeal. Among the city's attractions that the National Trust for Historic Preservation highlighted when it designated Natchez as one of its 2003 Dozen Distinctive Destinations is the diminutive and drastically underfunded Afro-American History and Culture Museum. The message was clear: Natchez is working toward what its city's tourism director called "a more balanced tourism product."[76]

Jim Crow may no longer live in the city's famed storybook mansions—African Americans are welcome to pay admission fees, just like white tourists—but the ghosts of its past are notoriously difficult to exorcise. By confining information about Natchez's deep and profitable association with slavery to a few selected sites, while home owners and docents across the city continue to venerate the antebellum past, one cannot escape the segregationist logic that prevents black and white perspectives from commingling in a shared historical landscape. Only when the memory embedded in southern landscapes breaks free from its racialized confinement, when we see whiteness as well as blackness at Forks of the Road and hear the voices of both black and white in the grandeur and oppression of Dunleith can Natchez be said to be living with its past, not haunted by it.

Acknowledgments

I want to thank Rich Schein for his careful reading of this chapter and helpful editorial suggestions, and Steve Wilson of the Harry Ransom Humanities Research Center for his assistance with the immense Selznick collection. I also wish to thank the many people in Natchez who shared their detailed knowledge of their city's historic landscape, especially Ser Seshs Ab Heter-C.M. Boxley, Jim Barnett, Alma Carpenter, Mimi Miller, Ora Frazier, Thom Rosenblum, Devereaux Slatter, and Cavett Taff.

Notes

1. Hodding Carter, *Lower Mississippi* (New York: Farrar and Rinehart, 1942).
2. Reid Smith and John Owens, *The Majesty of Natchez* (Montgomery, AL: Paddle Wheel Publishers, 1969), 6.
3. Anthony Walton, *Mississippi: An American Journey* (New York: Alfred A. Knopf, 1996), 47.
4. Some visitors never get beyond these first, rather unflattering, impressions. V.S. Naipaul is one who remained unimpressed by Natchez during his travels through the South. Natchez, he wrote, "was a wretched little town, steaming after rain on its 'bluff'—not very high—beside the muddy river." The Nobel Prize-winning author chose not to linger in Natchez, where he was repelled by "the jungle-sewer smell, the smell of the river, which was almost exactly like the jungle-sewer smell of Manaus, on the Amazon, in Brazil." It was only upon returning to the safe confines of his Jackson hotel that the ever-misanthropic Naipaul realized that "the rain, and the great heat, and my own ignorance of the beauties to look for, had kept me from the other wonder of Natchez. The river was altering its course; the bank at some place was being washed away; and some of the pretty old houses of the planter days were collapsing into the river." Naipaul, *A Turn in the South* (New York: Vintage Books, 1990), 218–20.
5. See http://www.nationaltrust.org/dozen_distinctive_destinations/2003/natchez.html (accessed October 16, 2004).
6. Smith and Owens, *Majesty of Natchez,* 6. Typical of guidebooks to Natchez, Smith and Owens's begins with a two-page photograph of a young woman in a white ball gown standing before the iron gates that open to an immense Greek Revival-style mansion, replete with two-story Corinthian columns encircling what is unmistakably a southern plantation estate.

7. "Come to Natchez, Where the Old South Still Lives and Where Shaded Highways and Ante-Bellum Homes Greet New and Old Friends, 'Garden Pilgrimage Week,' " 1932 poster, produced for the Natchez Garden Club. Copy located at the Historic Natchez Foundation, Natchez, Mississippi. After a trial run the year before, the first annual Natchez Pilgrimage took place in 1932.
8. Field notes from *A Southern Road to Freedom,* Natchez, Mississippi, March 15, 2001.
9. Mark Singer, "Never Surrender: The Sons of Confederate Veterans Have a Bad Day at the Mall," *The New Yorker,* May 14 2001, 52–57; Derek H. Alderman, "A Street Fit for a King: Naming Places and Commemoration in the American South," *Professional Geographer* 52, no. 4 (2000): 672–84; idem., "Street Names as Memorial Arenas: The Reputational Politics of Commemorating Martin Luther King, Jr., in a Georgia County," *Historical Geography* 20 (2002): 99–120; Jack Hitt, "Confederate Semiotics," *Nation* 264, no. 16, April 28, 1997, 11–15; Elizabeth Grace Hale, "We've Got to Get out of This Place," *Southern Cultures* 5, no. 1 (1999): 54–66; Jonathan I. Leib, "Heritage versus Hate: A Geographical Analysis of Georgia's Confederate Flag," *Southeastern Geographer* 35, no. 2 (1995): 37–57; and Gerald R. Webster and Jonathan I. Leib, "Whose South Is It Anyway? Race and the Confederate Battle Flag in South Carolina," *Political Geography* 20 (2002): 271–99. An excellent journalistic and travel account of the tensions that cut across the South's landscapes of memory and race is found in Tony Horowitz, *Confederates in the Attic: Dispatches from the Unfinished Civil War* (New York: Random House, 1998).
10. D.W. Meinig, "Symbolic Landscapes: Some Idealizations of American Communities," in *The Interpretation of Ordinary Landscapes,* ed. D.W. Meinig (New York: Oxford University Press, 1979), 164–92.
11. Margaret Mitchell, quoted in Aljean Harmetz, *On the Road to Tara: The Making of* Gone with the Wind (New York: Harry N. Abrams, 1996), 61.
12. Meinig, "Symbolic Landscapes," 164.
13. Alan Gowans, *Images of American Living: Four Centuries of Architecture and Furniture as Cultural Expression* (New York: Harper & Row, 1976), 281.
14. J.B. Jackson, *Discovering the Vernacular Landscape* (New Haven, CT: Yale University Press, 1984), 1.
15. John Michael Vlach, *Back of the Big House: The Architecture of Plantation Slavery* (Chapel Hill: University of North Carolina Press, 1993), 8. James C. Bonner wrote, "A great house of classic design … was a rare phenomenon, even among the more prosperous planters." Bonner, "Plantation Architecture of the Lower South on the Eve of the Civil War," *Journal of Southern History* 11, no. 3 (August 1945): 370–88.
16. Robert Gamble, "The White-Column Tradition: Classical Architecture and the Southern Mystique," *Southern Humanities Review* 12 (1978): 51–59. The varieties of American cultural landscapes during this time are described in D.W. Meinig, *The Shaping of America: A Geographical Perspective on 500*

Years of History, Volume 2, Continental America, 1800–1867 (New Haven, CT: Yale University Press, 1993).

17. Patrick Lee Lucas, "Realized Ideals: Grecian-Style Buildings as Metaphors for Democracy on the Trans-Appalachian Frontier" (Ph.D. diss., Michigan State University, 2002).
18. In Natchez, Haller Nutt's magnificent, and unfinished, showpiece Longwood exemplifies the trend in that city away from the Greek Revival style. An immensely wealthy planter and slave owner, Nutt conceived Longwood to be the grandest of the grand Natchez villas. The mansion was pure architectural fantasy: a three-tiered, cut-stone octagonal house "surmounted by a magnificent Persian dome" and ringed with minarets. Designed from a plate in Philadelphia architect Samuel Sloan's 1852 influential *The Model Architect,* the lacy "Moorish-Italianate" exterior woodwork finishes off a building in the so-called Oriental Villa style. The exterior was completed in the summer of 1861, several months after Mississippi seceded and one year before federal troops occupied Natchez and placed it under martial law. Samuel Sloan, *The Model Architect,* vol. 2 (1852; repr., New York: Dover Publications, 1980). For more on Longwood, see Randolph Delehanty and Van Jones Martin, *Classic Natchez* (Savannah: Martin-St. Martin, 1996), 144–47.
19. C. Reagles, "The Philosophy of Suburban Cottage Homes," *Southern Cultivator* 15 (1857): 325.
20. *Southern Cultivator* 14 (1856): 362.
21. *Soil of the South* 2 (1852): 253. It is also worth noting that, even when present, the neoclassical plantation landscape frequently failed to inspire. One antebellum traveler through the Deep South took special note of Natchez. Joseph Holt Ingraham recorded in that city "a huge colonnaded structure [which] struck our eyes with an imposing effect. It was the abode of one of the wealthiest planters of this state. … The grounds about this edifice were neglected, horses were grazing around the piazzas, over which were strewn saddles, whips, horse blankets, and the motley paraphernalia with which planters love to lumber their galleries." *The South-West by a Yankee, Volume 2* (New York: Harper and Brothers, 1835), 97–100.
22. Gamble, "The White-Column Tradition," 55. Margaret Mitchell, for one, recognized the disconnect between antebellum southern landscapes and the white-pillared past that the cinematic version of *Gone with the Wind* so successfully constructed. Indeed, even before the film's production, she complained about readers wanting a movie-set version of the South: "It's hard to make people believe that North Georgia [the setting for her novel] wasn't all white columns and singing darkies and magnolias." On the contrary, she was embarrassed at "finding myself included among those writers who picture the South as a land of white column mansions whose wealthy owners had thousands of slaves and drank thousands of juleps—if indeed it existed anywhere." Mitchell, quoted in Darden Asbury Pyron, "The Inner War of Southern History," in *Recasting*: Gone with the Wind *in American*

Culture, ed. Darden Asbury Pyron (Miami: University Presses of Florida, 1983), 185–201. Quote on page 186.

23. What is true for the symbolic landscapes of the New England village, Main Street, and the California suburb—the three landscapes identified by Meinig as America's symbolic spaces—is equally true for the plantation: it is the product of historical agents working to promote specific ideological, political, and economic agendas. A central point of this chapter is to demonstrate this claim. Joseph S. Wood, *The New England Village* (Baltimore: Johns Hopkins University Press, 1997); Richard V. Francaviglia, *Main Street Revisited: Time, Space, and Image Building in Small Town America* (Iowa City: University of Iowa Press, 1996); and Dolores Hayden, *Building Suburbia: Green Fields and Urban Growth, 1820–2000* (New York: Vintage, 2004). Two useful articles that advance this thesis for the South are John Radford, "Identity and Tradition in the Post-Civil War South," *Journal of Historical Geography* 18 (1992): 91–103; and Charles Reagan Wilson, "The Invention of Southern Tradition: The Writing and Ritualization of Southern History, 1880–1940," in *Rewriting the South: History and Fiction,* ed. Lothar Hönnighausen and V. Lerda (Tübingen: Francke, 1993), 3–21.
24. Edward L. Ayers, "Memory and the South," *Southern Cultures: A Journal of the Arts in the South* 2, no. 1 (1995): 5–8; David Blight, *Race and Reunion: The Civil War in American Memory, 1863–1915* (Cambridge, MA: Harvard University Press, 2001); Eric Foner, *Reconstruction: America's Unfinished Revolution, 1863–1877* (New York: Harper & Row, 1988); Grace Elizabeth Hale, *Making Whiteness: The Culture of Segregation in the South, 1890–1940* (New York: Vintage, 1998); Cecelia Elizabeth O'Leary, *To Die For: The Paradox of American Patriotism* (Princeton, NJ: Princeton University Press, 1999); Joel Williamson, *A Rage for Order: Black–White Relations in the American South since Emancipation* (New York: Oxford University Press, 1986).
25. C. Vann Woodward, *The Strange Career of Jim Crow* (New York: Oxford University Press, 1974 [1955]), xvi.
26. David Sansing, "Pilgrimage," in *Encyclopedia of Southern Culture,* ed. Charles Reagan Wilson and William Ferris (Chapel Hill: University of North Carolina Press, 1989), 700. That Natchez triggers recognition among African Americans, as well as whites, is nicely detailed in Anthony Walton's travel account through Mississippi: "I was going toward Natchez in search of that particular past, the plantation era. Perhaps the town's remnants of Old Mississippi had something to reveal about the beliefs and terrors that spawned the Mississippi I carried in my mind, the impressions and images, many violent, that I now wished to unravel" (Walton, *Mississippi,* 11).
27. David W. Blight, *Beyond the Battlefield: Race, Memory, and the American Civil War* (Amherst: University of Massachusetts Press, 2002); Gaines M. Foster, *Ghosts of the Confederacy: Defeat, the Lost Cause, and the Emergence of the New South* (New York: Oxford University Press, 1987); H.E. Gulley, "Women and the Lost Cause: Preserving a Confederate Identity in the American Deep South," *Journal of Historical Geography* 19 (1993): 125–41; Kirk Savage, *Standing Soldiers, Kneeling Slaves: Race, War and Monument*

in Nineteenth-Century America (Princeton, NJ: Princeton University Press, 1997); Charles Reagan Wilson, *Baptized in Blood: The Religion of the Lost Cause* (Athens: University of Georgia Press, 1980); John Winberry, " 'Lest We Forget': The Confederate Monument and the Southern Townscape," *Southeastern Geographer* 23, no. 2 (1983): 107–21. In 1890 the Natchez chapter of the United Daughters of the Confederacy, through the recently organized Confederate Memorial Association, erected the city's Confederate monument, with the wholehearted endorsement of the local newspaper. The *Natchez Democrat* (April 23, 1890) perfectly articulated the essence of the Lost Cause ideology when it noted,

> It must be remembered that [the monument] is a tribute of a people wasted and impoverished by war, whose country was overrun by invading armies and who after the storm of war was over, witnessed desolation on all sides—not only in their fields, but socially and politically—let it be remembered that out of and through this desolation, the people who are erecting this monument have never lost their gratitude, reverence, and devotion to the memory of their heroes who died a quarter of a century gone by but are now about to accomplish the crowning touch in enduring marble their undying sentiment.

In Natchez, for instance, the Confederate Memorial Association, which built the town's monument in 1890, disbanded in 1912. Melody Kubassek, "Ask Us Not to Forget: The Lost Cause in Natchez, Mississippi," *Southern Studies* 3, no. 3 (1992): 164.

28. Wilbur J. Cash, *The Mind of the South* (New York: Alfred A. Knopf, 1941), 124. Woodward again is prescient here as he describes the New South's "cult of archaism," "its nostalgic vision of the past. One of the most significant inventions of the New South was the 'Old South'—a new idea in the eighties, and a legend of incalculable potentialities." C. Vann Woodward, *Origins of the New South, 1877–1913* (Baton Rouge: Louisiana State University Press, 1951), 154–55. Following Woodward and as used by makers of the Natchez Pilgrimage, "Old South" refers not to antebellum history. Rather, it is shorthand for how southern whites have *imagined* the South to be before the Civil War; it is a paradigmatic example of how cultural memory creates what Edward Said called "imaginative geographies." Edward Said, *Orientalism* (London: Penguin, 1995). It should be added that white northerners joined in constructing this imaginative geography. Nina Silber, *Romance of Reunion: Northerners and the South, 1865–1900* (Chapel Hill: University of North Carolina Press, 1994). Hereafter, I will not use quotes around "Old South," but this distinct meaning remains.
29. For a useful overview of Cash, see James C. Cobb, "Does 'Mind' Still Matter? The South, the Nation, and *The Mind of the South*, 1941–1991," in *Redefining Southern Culture: Mind and Identity in the Modern South* (Athens: University of Georgia Press, 1999), 44–77. Francis Pendleton Gaines provided an early, incisive analysis of what I am calling the white-pillared past

in *The Southern Plantation: A Study in the Development and Accuracy of a Tradition* (New York: Columbia University Press, 1924).

30. Joel Chandler Harris, quoted in Gaines, *The Southern Plantation,* 13. Harris, it must be pointed out, was a far more complicated literary figure than many superficial treatments of him suggest. For two recent articles that persuasively demonstrate the subversive and even, at times, antiracist character of his work, see Wayne Mixon, "The Ultimate Irrelevance of Race: Joel Chandler Harris and Uncle Remus in Their Time," *Journal of Southern History* 56, no. 3 (August 1990): 457–80; and Robert Cochran, "Black Father: The Subversive Achievement of Joel Chandler Harris," *African American Review* 38, no. 1 (2004): 21–34.
31. Louis D. Rubin, "Uncle Remus and the Ubiquitous Rabbit," in *Critical Essays on Joel Chandler Harris,* ed. Bruce Bickley (Boston: G.K. Hall, 1981), 171.
32. Thomas Nelson Page, *The Old South: Essays Social and Political* (New York: Charles Scribner's Sons, 1894), 149. As for slavery, Page attempted to salve the sting by noting that "slavery, whatever its demerits, was not in its time the unmitigated evil it is fancied to have been. Its time has passed. … But to the negro it was salvation. It found him a savage and a cannibal and in two hundred years gave seven millions of his race a civilization, the only civilization it has had since the dawn of history" (Page, *The Old South,* 344). For more on Page, see Fred Arthur Bailey, "Thomas Nelson Page and the Patrician Cult of the Old South," *International Social Science Review* 72, no. 3/4 (1997): 110–21.
33. Fred Inglis, "Nation and Community: A Landscape and Its Morality," *Sociological Review* 25 (1977): 489–514. Quote on page 489. Roane Byrnes, one of the early garden club members most directly involved with both publicity and the pageant, believed that "it is not in our power to re-live our own lives, but we can, in some instances, *project ourselves into the past.* This may be done vicariously: by reading about the glories that are past, or hearing of them from one who has experienced that which we missed. However, a trip to Natchez offers anyone the opportunity of stepping into the Old South." Roane Byrnes, "Satisfy the Yearning for Yesterday—Go to Natchez" (typescript manuscript, Byrnes Collection, University of Mississippi Special Collections Library, Oxford, MS, n.d., Box 31, Folder 10).
34. Pierre Nora, "Between Memory and History: *Les Lieux de Memoire,*" *Representations* 26 (Spring 1989): 7–25. Quote on page 7.
35. Richard H. Schein, "Normative Dimensions of Landscape," in *Everyday America: Cultural Landscape Studies after J.B. Jackson,* ed. Chris Wilson and Paul Groth (Berkeley: University of California Press, 2003), 199–218.
36. Robert Gordon Pishel's *Natchez: Museum City of the Old South* (Tulsa: Magnolia, 1959) is merely one such guidebook to compare mansions like Melrose, where "the Old South reached its peak," with black neighborhoods that "meant violence and sin." *Natchez Democrat* quote from August 6, 1948.
37. Roane Byrnes, "Historic Natchez: Where the Old South Still Lingers" (unpublished manuscript, Byrnes Collection, University of Mississippi Special Collections Library, Oxford, Mississippi, n.d., Box 15, Folder 26).

38. National Park Service, "General Management Plan for Natchez National Historical Park," (Washington, D.C.: Government Printing Office, August 1991), 22. Not only were the region's planters extremely wealthy—one historian called the Natchez district "the richest principality in the domain of King Cotton in the decades leading up to the Civil War"—but they were also more likely to live in Natchez than on surrounding plantations. As the *New Orleans Crescent* reported, "Very few of these lords of the soil reside on their estates but have their residence across the river in Natchez." Michael Wayne, *The Reshaping of Plantation Society: The Natchez District, 1860–1880* (Baton Rouge: Louisiana State University Press, 1983). *New Orleans Crescent,* quoted in Wendell Garrett, "Foreword," in *Classic Natchez,* ed. Randolph Delehanty and Van Jones Martin (Savannah: Gold Coast Books, 1996), xi.
39. Describing these women as "elite" is not intended to denote only economic status. Many, in fact, were not rich, especially if they are compared to economic elites in other American cities during this period, and a good number struggled with considerable economic hardships in a chronically poor city. Rather, and much like the "elite" white women of Charleston, South Carolina, their status was marked by distinctive, local criteria: family name, marriage, and the economic and social status of their Mississippi slave-owning ancestors. For useful comparative studies, see W. Fitzhugh Brundage, "White Women and the Politics of Historical Memory in the New South, 1880–1920," in *Jumpin' Jim Crow: Southern Politics from Civil War to Civil Rights,* ed. Jane Dailey, Glenda Elizabeth Gilmore, and Bryant Simon (Princeton, NJ: Princeton University Press, 2000), 115–39; Robin Elisabeth Datal, "Southern Regionalism and Historic Preservation in Charleston, South Carolina, 1920–1940," *Journal of Historical Geography* 15, no. 2 (1990): 197–215; Richard R. Flores, *Remembering the Alamo: Memory, Modernity, and the Master Symbol* (Austin: University of Texas Press, 2002); Barbara J. Howe, "Women in Historic Preservation: The Legacy of Ann Pamela Cunningham," *Public Historian* 12 (1990): 31–61; Cynthia Mills and Pamela H. Simpson, eds., *Monuments to the Lost Cause: Women, Art, and the Landscapes of Southern Memory* (Knoxville: University of Tennessee Press, 2003); and Stephanie E. Yuhl, "Rich and Tender Remembering: Elite White Women and an Aesthetic Sense of Place in Charleston, 1920s and 1930s," in *Where These Memories Grow: History, Memory, and Southern Identity,* ed. W. Fitzhugh Brundage (Chapel Hill: University of North Carolina Press, 2000), 227–48. Economic data from Matt Volz, "Natchez Hopes Black Tourists Revamp Economy," *Associated Press State and Local Wire,* November 26, 2002 (accessed through Lexis-Nexis.com, November 30, 2004). National Register data is from http://www.cr.nps.gov/NR/research/ (accessed November 30, 2004).
40. Garden Club Natchez, "Eighth Annual Natchez Pilgrimage" (tourist brochure, Hill Memorial Special Collections Library, Louisiana State University, 1939). See also Katherine Boatner Blankenstein, "The Natchez Garden Club: A Brief History" (unpublished pamphlet, Archives Room of the Natchez Garden Club at Magnolia Hall, Natchez, MS, 1995).

41. Disagreements over how revenues from the Pilgrimage should be distributed led to an angry dispute within the Natchez Garden Club and to the creation, in 1937, of a rival organization known as the Pilgrimage Garden Club. For details on the so-called Battle of the Hoopskirts, see Hartnett T. Kane, *Natchez on the Mississippi* (New York: William Morrow, 1947); William H. Nicholas, "History Repeats in Old Natchez," *National Geographic*, February 1949, 181–208; and Blankenstein, "Natchez Garden Club."
42. Bette Barber, "Natchez' First Ladies: Katherine Grafton Miller and the Pilgrimage" (unpublished pamphlet, Mississippi Collection, First Regional Library, Hernando, MS, 1955); and Pilgrimage Garden Club, " 'Natchez, Where the Old South Still Lives': Announcement for an Illustrated Talk by Mrs. Balfour Miller (Originator of the Natchez Pilgrimage)" (typescript brochure, Katherine Miller Papers, Historic Natchez Foundation, Natchez, MS, n.d.).
43. "Natchez on the River," *House and Garden*, November 1939, 30–35. Harmetz, *On the Road to Tara*, 55. The David O. Selznick's papers, held at the Harry Ransom Humanities Research Center (HRC) at the University of Texas–Austin, contain more than a dozen folders of photographs documenting Selznick's research activities. Labeled "stairways," "doorways," "furniture," "dining room," "gardens," and so on, the vast majority of researched sites were in Natchez. On the back of one photograph showing the exterior of the Natchez home, "Rosalie," were the words: "Tara—this is the mansion" (HRC, Selznick papers, Box 3866, folders 1–13).
44. This is an important point taken up by many geographers writing on landscape and place from the perspective of language and text. More than a decade ago, Yi-Fu Tuan noted that only "the telling itself … has the power to endow a site with vibrant meaning." Tuan, "Language and the Making of Place: A Narrative-Descriptive Approach," *Annals of the Association of American Geographers* 81, no. 4 (1991): 684–96.
45. John Gunther, *Inside U.S.A.* (New York: Harper and Brothers, 1947), 803.
46. Shirley Wheatley, interview with the author, Natchez, MS, October 1999.
47. Mimi Miller, interview with the author, Natchez, MS, March 2000; Katherine Blankenstein, interview with the author, Natchez, MS, October 1999. "Natchez National Historical Park Comprehensive Plan" (Washington, D.C.: National Park Service, 1991), 3–4. William Johnson, *William Johnson's Natchez: The Antebellum Diary of a Free Negro*, ed. William Ransom Hogan and Edwin Adams Davis (Baton Rouge: Louisiana State University Press, 1951).
48. Tim Cresswell, *In Place/Out of Place: Geography, Ideology, and Transgression* (Minneapolis: University of Minnesota Press, 1996).
49. Allison Davis, Burleigh B. Gardner, and Mary B. Gardner, *Deep South: A Social Anthropological Study of Caste and Class* (Chicago: University of Chicago Press, 1941), 193.
50. Lansdowne home tour, March 2000; The Elms home tour, March 2000; The Briars home tour, March 2000; Green Leaves home tour, April 1998.

51. Rosalie home tour, April 1999; Montaigne home tour, April 1999. See also "Spring Pilgrimage," *Natchez Democrat,* March 10, 1999, special edition.
52. D'Evereaux home tour, October 1999; Lansdowne home tour, March 2000; Routhland home tour, April 1999; Linden home tour, March 2001; and Oakland home tour, March 2000.
53. The sole exception can be found at Melrose. Operated by the National Park Service and not part of the pilgrimage "old home" tours, Melrose stands apart in its efforts to include slavery as part of its interpretation. Melrose home tour, October 1999. As controversy at Colonial Williamsburg makes clear, however, simply including slavery as part of the tour narrative by no means solves all problems of interpretation. Richard Handler and Eric Gable, *The New History in an Old Museum: Creating the Past at Colonial Williamsburg* (Durham, NC: Duke University Press, 1997). For an informative, comparative analysis of plantation museum interpretations in Virginia, Georgia, and Louisiana, see Jennifer L. Eichstedt and Stephen Small, *Representations of Slavery: Race and Ideology in Southern Plantation Museums* (Washington, D.C.: Smithsonian Institution Press, 2002).
54. Monmouth home tour, October 1999.
55. Longwood home tour, March 2000. Information on Haller Nutt is from Michael Wayne, *The Reshaping of Plantation Society: The Natchez District, 1860–1880* (Baton Rouge: Louisiana State University Press, 1983), 9–10. See also Anthony E. Kaye, "The Personality of Power: The Ideology of Slaves in the Natchez District and the Delta of Mississippi, 1830–1865" (Ph.D. diss., Columbia University, 1999).
56. Georgia Wilson Newell and Charles Cromartie Compton, *Natchez and the Pilgrimage* (Kingsport, TN: Southern Publishers, 1935), 24–25.
57. Nell Savage Mahoney, "The Melody Lingers On," *Country Life,* April 1935, 10–15. That such servant roles were meant to be more akin to landscape—features to be viewed from a distance—than to tour guides was demonstrated to John Gunther, who learned that the man serving him cocktails had read one of his books. The incident "enrage[d] several people present, and puzzle[d] others to the point of consternation," because "it was literally unthinkable to [his white hostess] that this evidence of mild literacy by a black underling could be possible" (Gunther, *Inside U.S.A.*, 803).
58. Edith Wyatt Moore, "The Confederate Pageant and Ball: Eleventh Annual Pageant of the Original Natchez Garden Club" (March 7–April 7, 1942, Box 1, Natchez Garden Club Records, Mississippi Department of Archives and History, Jackson, MS). According to the logic of the early Confederate pageants, "Steamboats, cotton pickers and pickaninnies … formed a triumvirate lending enchantment to old plantation days. The rich, barbaric strains of the colored singer mingled strangely with the clamor of field activities and often drowned its raucous sounds." The term *Second Reconstruction* is C. Vann Woodward's and refers to the years of the civil rights movement in which black activism reached it peak and when overt southern white resistance—if not racism—began to subside. C. Vann Woodward, "Look Away, Look Away," in *The Burden of Southern History* (Baton Rouge: Louisiana State

University Press, 1993), 235–64. Marsha Colson, interview with the author, Natchez, MS, March 2000; Mary Toles, interview with the author, Natchez, MS, October 1999.

59. Ora Frazier, interview with the author, Natchez, MS, March 2001. See also Jack E. Davis, *Race against Time: Culture and Separation in Natchez since 1930* (Baton Rouge: Louisiana State University Press, 2001).
60. The "massive resistance" of Pilgrimage leaders to civil rights workers is detailed in the 1964 and 1965 Pilgrimage Garden Club Meeting Minutes, Pilgrimage Garden Club Records, Stanton Hall, Natchez, MS, and summarized in Steven Hoelscher, "Making Place, Making Race: Performances of Whiteness in the Jim Crow South," *Annals of the Association of American Geographers* 93, no. 3 (2003): 657–86. African Americans were granted admission to the museum-home tours in 1967.
61. Pilgrimage brochure, 1947, Roll 8, Natchez Garden Club Records, Mississippi Department of Archives and History, Jackson, MS.
62. Jim Yardley, "Black History, Civil Rights Luring Tourists to the South," *Atlanta Journal Constitution,* November 13, 1992, A1, A6; Gerrie Ferris, "Around the South in Search of the Past: Black Tourism Industry Thrives in '90s Search for Black Heritage," *Atlanta Journal Constitution,* June 30, 1996, D6; Reginald Owens, "African American Tourism Is Growing Source of Pride and Business: On the Heritage Trail," *New Orleans Times-Picayune,* July 16, 1995, F1; and Russ Bynum, "Savannah Cashes in on Boom in Black Tourism," *Newsday,* 10 June 2001, E15. Attracting black tourists by developing black historical sites is becoming an acknowledged goal of Natchez city officials. The black heritage market is "relatively untapped," said Natchez mayor F.L. Hank Smith. "There is a side of the story that needs to be told. There is a growing segment of the tourism market that wants to see the story" (Volz, "Natchez Hopes Black Tourists Revamp Economy").
63. Selma Mackel Harris, interview with the author, Natchez, MS, March 2001; Ozelle Fisher, interview with the author, Natchez, MS, October 1999; Ora Frazier, interview with the author, Natchez, MS, March 2000. In 2001, and as a result of those very antagonisms, the pageant was renamed "Historic Natchez Pageant."
64. Here and later, information on *A Southern Road to Freedom* comes from observations of performances in April 2000 and March 2001.
65. Bertolt Brecht, "Theatre for Pleasure or Theatre for Instruction," in *Brecht on Theater: The Development of an Aesthetic,* ed. and trans. John Willett (New York: Hill and Wang, 1964), 71. *Verfremdungseffekt,* or defamiliarization, is Brecht's term for a performance that makes the familiar strange and hence one that pushes audience members to forge new understandings of complex realities; it is a concept employed to great effect in *A Southern Road to Freedom.* Despite *Southern Road*'s success—the gospel performance is the only Pilgrimage event to enjoy increasing attendance over the past several years—it remains marginalized as "something else a tourist can do while in town." It receives far less attention from the Natchez publicity machine than other attractions, and in 2000 the Mississippi State tourism guide failed even

to mention the event. Moreover, as provocative as the performance is, by its very nature *A Southern Road* is ephemeral and without a spatial anchor; it lacks a landscape of memory.

66. Ser Seshs Ab Heter-C.M. Boxley, interview with the author, Natchez, MS, December 2004. The writer Joseph Holt Ingraham traveled through Natchez in the early 1830s and provides the most vivid account of the Forks of the Road slave market site. Describing the slave dwellings at the site, Ingraham wrote, "The contrast between the miserable buildings and their squalid occupants, and the rich woodlands beyond them on either side, among whose noble trees rose the white columns and lofty roofs of elegant villas, was certainly very great, but far from agreeable" (*The South-West, Volume 2,* see especially pp. 190–205, 238–47; quote on page 190).
67. Jim Barnett and H. Clark Burkett, "The Forks of the Road Slave Market Site at Natchez," *Journal of Mississippi History* 63, no. 3 (2001): 168–87; quote on page 178. Ingraham estimated that in 1834 alone more than four thousand enslaved people were brought into Mississippi and one-third of them were sold in the Natchez market (Ingraham, *The South-West, Volume 2,* 244).
68. Kenneth Foote, *Shadowed Ground: America's Landscapes of Violence and Tragedy,* rev. ed. (Austin: University of Texas Press, 2003), 24, 174–213.
69. Boxley interview. "Many of the people who come to Natchez," Ser Boxley explained to a reporter from Oakland, "see the mansions as the great memory of 'Gone with the Wind' and Scarlett O'Hara, but do not consider that it was mostly Africans who came to America in chains who helped build and maintain these mansions." Charles Aikens, "Clifford Boxley Researches Natchez Slave History to Tell Story of Its Mansion Builders," *Oakland Post,* October 17, 1999, 7. Details on Ser Seshs Ab Heter-Clifford M. Boxley and his experiences of growing up in Jim Crow Natchez are detailed in the *History of Jim Crow Project:* http://www.jimcrowhistory.org/resources/narratives/Clifford_Boxley.htm.
70. Sherri Williams, "Forks of the Road Pain Inspired Quest," *Clarion-Ledger* (Jackson, MS), February 29, 2000. The committee raised $1,400 for the marker. "Mississippi Juneteenth Celebration Calendar," www.juneteenth.com/4mississippi_us.htm (accessed November 30, 2004).
71. Thom Rosenblum, interview with the author, Natchez, MS, December 2004; Jim Barnett, interview with the author, Natchez, MS, December 2004; Cavitt Taft, interview with the author, Jackson, MS, December 2004; Matt Volz, "Slave Market Site Reflects Racial Fork," *Clarion-Ledger* (Jackson, MS), December 29, 2002; and Volz, "Slave Market Story Gets Complicated," *Biloxi Sun Herald,* December 21, 2003. Many of the details are listed on Ser Seshs Ab Heter-Clifford M. Boxley's website devoted to the Forks of the Road site: www.bjmjr.com/forks_roads/index.htm.
72. Rosenblum interview.
73. Boxley interview. The conflict that Ser Boxley described lies at the center of what Dolores Hayden called the "power of place"—a conflict over the meaning and value of historic landscapes that frequently pits questions

of aesthetics against those of social relations. Hayden, *The Power of Place: Urban Landscapes as Public History* (Cambridge, MA: M.I.T. Press, 1995).

74. Mike Brunker, "Race, Politics, and the Evolving South: A Black Mayor, 130 Years Later," August 17, 2004, MSNBC.com, www.msnbc.com/id/5676325 (accessed October 13, 2004).
75. Patricia Leigh Brown, "New Signpost at Slavery's Crossroads," *New York Times*, December 16, 2004. Cavett Taff, interview with the author, Jackson, MS, December 2004.
76. See http://www.nationaltrust.org/dozen_distinctive_destinations/2003/natchez.html (accessed October 16, 2004). Walter Tipton, Natchez tourism director, quoted in ibid.

CHAPTER 4

Seeing Hampton Plantation: Race and Gender in a South Carolina Heritage Landscape

SAMUEL F. DENNIS JR.

Getting to Hampton Plantation

To get to Hampton Plantation State Historic Site, you drive north from Charleston, South Carolina, along U.S. Highway 17 for about half an hour. On a sunny August morning, all four lanes of the "coastal highway" are crowded with cars. It is the height of the tourist season, and the oceanfront hotels and resort condominiums in Myrtle Beach are overflowing. But Myrtle Beach is yet another fifty miles north and lies at the center of a sprawling region of golf course developments, strip malls, and amusement parks. This stretch of highway, however, runs dead-straight through miles of unbroken pine forest. On your left is the Francis Marion National Forest, and on your right are the salt marshes and sea islands of the Cape Romain National Wildlife Refuge. No oceanfront hotels. No golf resorts. No country-music halls. You are experiencing the longest stretch of undeveloped coastal land in South Carolina.

Just north of the fishing village of McClellanville you make a left onto Rutledge Road, a two-lane asphalt road crowded on both sides by tall pines that encroach so close to the highway shoulder that their canopies almost touch overhead. This road heads directly into the heart of the national forest, and yours is the only car on the highway now. The empty road encourages speeding, and you almost miss the entrance to the state park.

Hitting the brakes hard, you ease your car off the asphalt and onto the soft white sand of the long entrance drive. The tree canopy closes overhead; the driveway is a ribbon of white in the dim light. The sand is deep, and your car pitches and yaws on the uneven surface. You get the sensation of piloting a boat through gently rolling waters. Eventually the drive straightens out, but just as you become aware of the strong axis bearing you toward a brightening opening in the forest—this must be the formal plantation entrance you have been expecting—a gate appears, and you are directed around the left side of the plantation grounds toward a parking area next to an open field of grass. Although it looks like it might accommodate a hundred cars, there is only one vehicle parked in the lot on this day.

More than anything else, the drive to Hampton Plantation conveys a powerful sense of a world apart, and in many ways this isolation is no different today than in the period before the Civil War. The slaveholding families of the Carolina coastal region—the so-called Lowcountry—were only seasonal residents on rice plantations such as Hampton, if they resided there at all. Most moved between townhouses in cities such as Charleston, cottages in the pine forests of the interior coastal plain, and beachfront houses on the coastal barrier islands. This seasonal migration pattern held for postbellum planter families as well.

The last owner of Hampton Plantation, the writer and South Carolina poet laureate Archibald Rutledge (1883–1973), grew up in such a household. He was born at his family's summer place in McClellanville, although he often wrote that he was born on Hampton Plantation.[1] As a teen in the late nineteenth century, Archibald and his brother stayed in the village during the summer, while his father, the plantation master, escaped the heat, humidity, and mosquitoes by heading to the cool of the North Carolina mountains. As it was for generations of white slaveholding families before emancipation, the plantation landscape was best avoided during the summer months. According to Elizabeth Allston Pringle, her family would not leave their beach cottage at Pawley's Island and return to the rice plantation before a "black frost" had occurred, usually in November—"three white frosts make a black frost," she wrote.[2] From the earliest European settlement of the Carolina coast, those who could afford to do so avoided the malaria and yellow fever prevalent in the Lowcountry during the rice-growing season. Of course, enslaved laborers had no such freedom of movement, and for this reason the Carolina Lowcountry was from colonial times a region in which African American cultural practices were predominant.[3]

Part of the attraction of low-country heritage landscapes such as Hampton Plantation lies rooted in their association with African cultural survivals, including Gullah, the Creole language of low-country African Americans. This interest first peaked during the late 1930s when the Federal Writers' Project, part of the Depression-era Work Projects Administration, documented oral histories of former slaves. The Georgia series in particular focused on African cultural survivals among the coastal African American population.[4] The isolation that preserved the Creole culture among low-country African Americans also prompted romantic visions of an idealized Old South culture in the imagination of white visitors and residents alike. Julia Peterkin, South Carolina low-country resident and novelist, wrote in the 1930s, "Some of the charm that made the life of the old South glamorous still lingers on a few plantations that have been so cut off from the outside world by rivers, wide swamps, and lack of roads that they are still undisturbed by the restless present."[5]

On another August day Archibald Rutledge returned to Hampton Plantation after more than three decades teaching English at Mercersburg Academy in Pennsylvania. He acknowledged the effect of isolation on the plantation landscape (see Figure 4.1).

Figure 4.1 View of Hampton Plantation. Little has changed since Rutledge's return in 1937. Photographed by author, 2004.

> When in August, 1937, I returned to Hampton, the road was almost closed by a lush growth of grass and briars and bushes. When I came to where the gate used to be, I could hardly see the house for the tall weeds and the taller bushes. It was as if the blessing of fecundity had been laid on everything natural, and on everything human, the curse of decay.[6]

For Rutledge, as with other descendents of low-country slaveholding families, the rice plantation had always been a world apart—a lonely place far from civilization, surrounded by wild nature, visited infrequently for hunting trips, perhaps, during holidays and summer vacations. On the other hand, of course, this landscape had been inhabited by African American families for generations. Isolation was then, and remains today, a relative phenomenon. What was romantic—even exotic—to whites was to others simply home.

Seeing Hampton Plantation

There are at least two ways to see the cultural landscape at Hampton.[7] The first emphasizes long-standing links between African American families and the land—first through their enslavement by Rutledge's ancestors but later through the world these same families made as freed men and women. Like on other plantations throughout the region, freedmen and freedwomen stayed on at Hampton after the end of slavery, in large measure, because they had access to land.[8] Throughout the Lowcountry, African American communities built their postemancipation lives on the informal economy that emerged during slavery as a network of relations beyond planter control, one that connected provision gardens on the plantation to urban markets by means of the many waterways along the coast.[9] The intimate link between postbellum African American communities as social units and the plantation landscapes in which they lived was so strong that in some cases the simple movement of particular households redefined plantation boundaries.[10]

A second way of seeing the plantation landscape emphasizes the planter's view of the land and includes a geographical imagination connected to notions of private property.[11] A dominant planter perspective is bound to the bird's-eye view of landscape paintings and the aerial views of maps and land surveys.[12] Rutledge wrote about his pleasure in possessing a map of Hampton produced for one of his ancestors that showed the different rice fields in contrasting watercolors.[13] The connection between landscape and social status was produced in a number of ways. One was simply having one's family name linked to the land. In his 1825 atlas, for example, Robert

Mills labeled plantations with family names.[14] Landscape and social status were also linked through the formal layout of the plantation. One rationale for spatial organization in the plantation landscape can be explained functionally, such as the location of slave quarters near the overseer's residence for improved surveillance. But for slaveholding white families, the plantation was above all else a processional landscape through which visitors moved during social rituals.[15] The plantation landscape served symbolic functions meant to highlight the power and wealth of the slaveholding family.[16] Long entrance drives, for example, often carried visitors past the slave quarters, the number of which offered some measure of planter wealth in human property.

At the Hampton Plantation State Historic Site, the visible scene is dominated by the plantation "big house," although the site encompasses several mostly modern outbuildings, including a restroom facility and a picnic shelter. An old brick chimney stands where Prince Alston, a son of Hampton Plantation slaves, built a home from salvaged materials for himself and his wife Sue (see Figure 4.2). Alston (1883–1928) was one of Rutledge's hunting partners, a childhood companion and frequent subject of his plantation hunting stories. Alston's wife, Sue, likewise appeared in many of Rutledge's stories, and when Hampton was sold to the state of South Carolina in 1971, Rutledge stipulated that she be allowed to live on Hampton as long as she wished. The ruins of her house are interpreted through a series of photographs and explanatory text. The sign reads in part,

Figure 4.2 Ruins of Prince and Sue Alston's home. Photographed by author, 2004.

> The chimney in front of you is all that remains of the Alston tenant house. It is one of the last standing structural reminders of African-Americans at Hampton Plantation and a testament to the resourcefulness of Prince and Sue Alston. The Alstons and other descendents stayed on at Hampton Plantation long after rice—the once valuable "Carolina Gold"—was no longer a profitable crop in the South Carolina Lowcountry.

Other than the big house, not much else remains of the working rice plantation (see Figure 4.3). There are no slave quarters, there is no overseer's house or whipping post, and there are not any reminders of the postbellum African American community that numbered around seventy-five when Rutledge returned in 1937.[17] In fact, the only visible reminder of the enslaved labor that produced the plantation landscape lies at the end of a pathway leading from the big house to Wambaw Creek a short distance away. Here one can look across the water to the land beyond (see Figure 4.4), but without the sign explaining that the scene was once a rice field, it would have no particular meaning for the visitor.

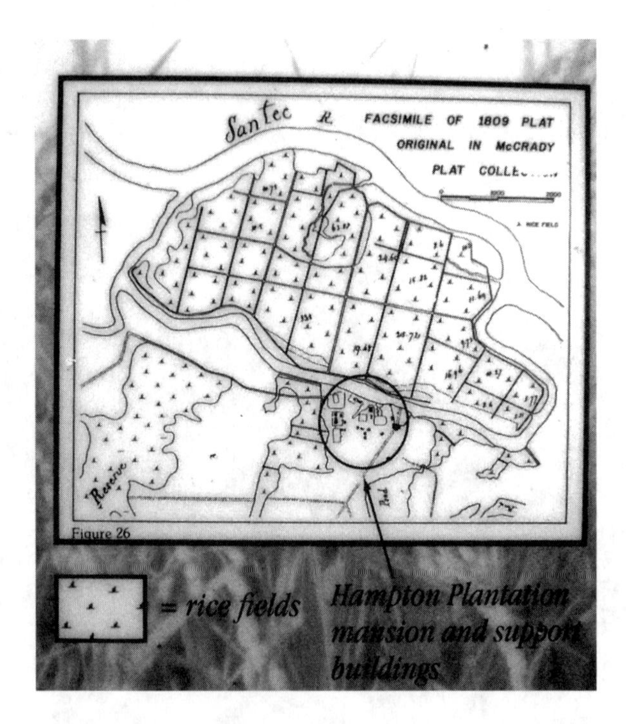

Figure 4.3 Map of Hampton, circa 1809. Note the rice fields north of the plantation home, across Wambaw Creek. Detail from site display, photographed by author, 2004.

Figure 4.4 Former rice fields north of Wambaw Creek. Photographed by author, 2004.

The present-day visitor to Hampton Plantation also does not see the hundreds of acres of rice fields originally created, cultivated, and maintained by enslaved Africans and African Americans. Rice growing produced the material wealth of the white slaveholding families and defined the Lowcountry as a distinct region of the southeastern United States. The Lowcountry refers to the area of the lower coastal plain in which freshwater rivers experience the rise and fall of the tides, a condition necessary for lowland rice cultivation. Floodplains along the rivers and midriver islands were stripped of timber and converted to rice fields, protected by an enclosure of embankments separating the fields from the high tides that would otherwise flood the land. Multiple rice fields were surrounded by the highest embankments, while individual fields were segregated through a series of crossbanks. A system of ditches and trunks (control gates) distributed fresh water to the individual fields, which could be flooded on the rising tide and drained after a period of time on the ebbing tide.[18] Dikes and ditches are still visible on aerial images of the Lowcountry (see Figure 4.5), although they are difficult to recognize from land or water (see Figure 4.6).

Like other heritage sites in the Carolina Lowcountry, the landscape at Hampton Plantation is made to tell the story of the white slaveholding family and its descendents, with special attention given to the plantation master. A study of the narratives employed at plantation heritage sites in Louisiana, Georgia, and Virginia found that the stories centered on whites, elites, and males at the expense of the labor and lives

Figure 4.5 Aerial infrared image of Georgetown, South Carolina and the Wacamaw River. Note the ditches (straight lines) separating the former rice fields still visible in the image. Source: Detail from poster created by the North Inlet–Winyah Bay National Estuarine Research Reserve using near-infrared images from the National Aerial Photography Program.

of enslaved African American families.[19] The study authors argued that the public stories presented at these heritage sites construct and maintain white racial identities that are linked, through a narrow and partial reading of history, to white pride in the foundations of American democracy. Plantation sites employ a variety of representational strategies to frame their landscape narratives. The most frequently used strategy is "symbolic annihilation and erasure," which describes the removal, both materially and discursively, of any trace of African American presence. Another strategy seeks to trivialize the conditions and contributions of enslaved laborers by presenting slavery as a benign, even benevolent, institution and by presenting white masters as kindhearted protectors and providers. Hampton Plantation is a relic *visual* landscape, rather than a working, agricultural landscape. The focus is on the big house rather than on the rice fields. The visitor is presented with a narrative that frames the plantation landscape and the African Americans who shaped and inhabited it through the story of Rutledge's life.

Figure 4.6 Photograph from abandoned bridge (the broken line north of the active bridge in aerial image). Notice that what is easily seen in the aerial image is difficult to detect on the ground. Photographed by author, 2004.

The South Carolina Poet Laureate Returns

Rutledge was a prolific writer who perhaps was the most widely read low-country writer of his time. Others included Julia Peterkin, who was awarded the Pulitzer Prize in 1929 for her novel *Scarlet Sister Mary*, and DuBose Heyward, best known for his collaboration with George Gershwin on *Porgy and Bess* (based on Heyward's 1925 novel *Porgy*). Although both were more critically acclaimed, neither Peterkin nor Heyward matched Rutledge in sheer productivity or in wide dissemination through popular national magazines. Rutledge was named South Carolina's first poet laureate in 1934 and was best known for his plantation stories, which included hunting tales, works in natural history, and anecdotes about the African American tenants living at Hampton. His short fiction appeared in dozens of national magazines, including the *Saturday Evening Post*, *Harpers*, *Field and Stream*, and *Scribner's*. He published more than thirty collections of short prose during his career, in addition to the many collections of his poetry.[20] In 1930 Rutledge was awarded the Burroughs Medal for natural history writing, an honor he shared with other writers such as Aldo Leopold and Rachel Carson.

When Rutledge left his teaching position at Mercersburg Academy in Pennsylvania, he retired to South Carolina and planned to operate Hampton as a living plantation museum, while spending his leisure time hunting and fishing in the surrounding countryside. His dream was to restore the big house with the original furnishings and "beautify the grounds so that the place had a loveliness as an attraction as well as its deep historic significance."[21] The big house, the landscaped grounds—even the furniture—figured prominently in many of his plantation stories. These were often used to link Hampton with other prominent low-country planter families, important historical events, or national heroes. For example, Rutledge used the large moss-draped oak tree in the front garden to narrate a story of George Washington's 1791 tour of the south. During his brief stop at Hampton, so the story goes, the president was asked whether the tree's placement interfered with the symmetry of the front portico and, if so, should the tree be removed. "Let the tree stand," Washington is reputed to have said. To this day, the "Washington Oak" features prominently in the landscape narrative at Hampton. Rutledge refers to other historical figures as well, in particular his distant relative Edward Rutledge, a signer of the Declaration of Independence. These examples illustrate the rhetorical strategy used to link the white plantation family to the nascent American democracy. In addition, keeping the focus on national history served to invite white visitors to share in the pride felt by the plantation family, while at the same time deflecting attention away from the institution of slavery generally and the lives of Hampton's enslaved African Americans in particular.

In his restoration work at Hampton, as in his fiction, Rutledge was concerned to highlight the historic narratives of the plantation through the architectural evolution of the big house. The period of its original construction is used to tell the story of French Huguenot settlement in the Lowcountry. Each addition or architectural embellishment provided an opportunity to situate the family in the flow of American history. The rhetorical strategy of connecting the plantation family to important currents in American history through the story of the big house was most clearly articulated in Rutledge's 1940 *Home by the River* and later expanded as the primary narrative presented by the State Historic Site.[22] Rutledge was equally interested in restoring the plantation landscape in ways that presented the house in the most beautiful of settings. "Certain writers have invested plantation life with glamour," he wrote, "I believe it had that glamour. Other writers have insisted that plantation glamour was a myth. I think they are wrong."[23] The tranquil landscape surrounding the plantation house, together with the absence of material references to slavery, clearly illustrates Rutledge's desire to present Hampton to visitors as a romantic and untroubled allusion to the glamour of the Old South.

Rutledge's 1937 retirement to Hampton signaled something beyond his simple homecoming. He was self-consciously stepping into the role of the plantation master, the rustic but cultured host he would play for visitors up until he permanently moved to Spartanburg following a hip injury in 1967.[24] "It's a gay cavalcade that visits Hampton," he wrote, "people from all states come to see what America once was."[25] Many of those visitors were his neighbors, recent owners, and seasonal residents of other low-country rice plantations. Hampton was one of hundreds of former low-country plantations reinhabited between the 1890s and the 1930s. A few were purchased by hunting clubs, but most were bought up by wealthy businessmen from the northeast for use as winter hunting retreats.[26] Another low-country rice planter and writer, Elizabeth Allston Pringle, complained, "One multimillionaire ... who bought a very fine rice plantation, Prospect Hill, formerly property of William Allston, has some fields planted in rice every year, simply for the ducks, the grain not being harvested at all, but left to attract the flocks to settle down and stay there, ready for the sportsman's gun."[27] This new planter class, its wealth produced through industrial development rather than land rents or agriculture, was interested in the social imagery of the landed aristocracy. Raymond Williams noted that, in England, the newly wealthy industrialists of the nineteenth century adopted a "rural mode of display" centered on the rituals of riding and hunting.[28] The Lowcountry was little different; it was a world of leisure that linked hunting on former rice plantations to polo matches in Aiken, horse racing in Camden, and golf tournaments in Augusta.

Like the antebellum plantation families before them, the new planters were only seasonal residents in the Lowcountry. Still, they brought needed cash into the regional economy. Rutledge was able to attend college, for example, on funds his father generated leasing hunting rights on Hampton Plantation. More important, however, these seasonal residents constituted an eager audience for plantation narratives, and in this Rutledge excelled. "The very wealthy owners of great estates," he wrote, "have been most gracious to me, and often come informally to visit this rustic native. And I am privileged to go right into their exquisite homes with my hunting clothes on."[29]

To properly perform the role of plantation master required a supporting cast. Rutledge acknowledged upon his return in 1937 how forlorn the plantation grounds appeared. "Yet, when I drove up to the house," he wrote in the opening of *Home by the River,* "there were seven Negroes, a loyal if tattered company, to meet me. They said little, but I could feel their affection and eagerness to start."[30] These few sentences begin to articulate the ideology of planter paternalism—the myth that evolved during the early nineteenth century among slaveholding families that imagined enslaved African

Americans as childlike and incapable of taking care of their own affairs without the white plantation master's protective care. In this, and many of his other writings, Rutledge reinscribed the discourse of paternalism, and that discourse continues today, however subtly, to dominate the interpretation of the Hampton Plantation landscape. The discourse of paternalism continues to construct racial and gender categories through the dominant white cultural memory of low-country plantation life and landscape.

Plantation Paternalism

Planter paternalism was a self-serving ideology that emerged in the late eighteenth and early nineteenth centuries among white slaveholding families who, Betty Wood argued, came to "believe the image of themselves as benevolent, paternalistic masters who tended to all the material and spiritual needs of their bondpeople."[31] The ideology was expressed in the oft-repeated phrase "my family, black and white" and was revived during the early twentieth century through a variety of media. Regional writers and artists reinvented a plantation past where enslaved African American men and women lived contented lives under the benevolent care of the white plantation master. Planter paternalism described a set of reciprocal obligations between white planters and enslaved black laborers as members of the same household. Rutledge had sharecropping arrangements with the tenants on Hampton—he referred to these as "my Negroes"—in which he furnished seeds, implements, and a plow team in return for 15 percent of the crop. He did not seek profit in this endeavor; his was a life of leisure, hunting in the woods surrounding Hampton or working on the restoration project. In his many plantation stories, Rutledge occasionally is asked to pass judgment in petty disputes, and he often hands out a few dollars for this and that, offering the reader a sense of his obligations to his plantation family. In this way, Rutledge masks the material dominance he enjoys by filtering plantation relations through the lens of paternalism.

As Rutledge did in his writing, low-country artist Elizabeth Ravenel Huger Smith (1876–1958) worked to visualize the romantic landscapes of rice plantations, giving form through her watercolor paintings to the ideology of planter paternalism. These were originally planned to illustrate a book she and her father, D.E. Huger Smith, had "intended as a laurel wreath for that great civilization, of the rice-planting era in South Carolina."[32] Although her father passed away just as they began work on the book, the final version, written by Herbert Ravenel Sass, included her father's nostalgic reminiscences about life on a rice plantation in the 1850s.[33] Smith's images depict enslaved plantation workers in landscapes that seem timeless and idyllic. Women working in the rice fields, for example, are beautifully (and anonymously) depicted. They seem to be leisurely socializing as much as working. One painting in particular, *Sunday Morning at the Great House,* shows the white slaveholding family receiving well-dressed slaves on a Sunday afternoon at the big house. The white plantation master seems to be presiding over a very large family gathering, the people showing nothing but respect for each other. Taken together these thirty paintings suggest a familial bond between enslaver and enslaved, but even a cursory reading of the narratives of former low-country slaves recorded during the 1930s reveals the brutality with which these relations were negotiated.[34]

Other white voices from the 1920s and 1930s, including Julia Peterkin and Dubose Heyward, were more sympathetic to the lived experience of low-country African Americans. Peterkin highlighted the extent to which planter paternalism served white male interests in her characterization of low-country plantation life in the 1930s.

> Servants are plentiful and cheap on the plantations, and it is taken for granted that every Big House family, poor or prosperous, is entitled to service. A gentleman may confess and lament his poverty, but he is not considered extravagant if a "body servant" brings his coffee before he gets out of bed in the mornings, presses his clothes, shines his boots, saddles his horse and attends to his hounds and bird dogs. The body servant's duties may also include ... keeping his [employer's] spirits cheered whenever he gets down in the heart.[35]

Rutledge materially lived the character of the plantation master, performing a somewhat self-conscious living tribute to a romanticized plantation heritage. His stories encoded the landscape at Hampton with racial and gender meaning, in large measure through the associations of gender and race with particular visible relations to the landscape, through either female labor in the fields or male labor in the woods. Proper social relations on the plantation, as in the region generally, were thus communicated through plantation and hunting stories that translated the ideology of paternalism into the visible scenes of everyday living in the Lowcountry.

White, Black, Man, Woman

More than anything else, planter paternalism worked to naturalize racial and gender categories. The narratives of paternalism presented the unequal power relations between the races as a simple expression of the natural order of things rather than an expression of the historically contingent social conditions of slavery. The planter way of seeing landscape imposed an ideological order on the visible plantation scene as well as on the mix of public and private spaces beyond the plantation. In part, this was a reaction to the informal slave economy that emerged in the Lowcountry during the early nineteenth century and provided enslaved low-country African Americans a degree of freedom not shared by their counterparts in the cotton- and tobacco-growing regions. Rutledge, however, relied on this ideological legacy to explain racialized social relations in the postbellum plantation landscape.

Rutledge believed that low-country African Americans had a special, indeed mysterious, relationship with the natural world. "I came to love and study nature," he wrote, "Negroes were my daily companions, and from them I learned something of the ways of the natural world."[36] Rutledge did most of this learning during the ritual of the hunt, where for years he was accompanied by Prince Alston. Prince served Rutledge in these outings as guide, oarsman, driver, and tracker. In Rutledge's view, African Americans like Prince possessed a "certain animal intelligence," a connection to animals as "brothers and sisters," and a natural ability to interpret meaning in nature, or as Rutledge put it, to "deeply read in the oracles of God."[37] Prince in particular possessed an "almost occult understanding of all creatures, wild or tame," wrote Rutledge. "The savage bushmen of the forests of equatorial regions of Africa," Rutledge continued, "bring to their pursuit of game a certain aboriginal directness of comprehension that is beyond the power of white man to employ naturally or even to imitate. Prince had this same kind of power."[38] Julia Peterkin wrote in a similar vein, although she presented this understanding of nature less as a perceived racial trait than as a product of the social conditions under which low-country African Americans lived. "On large plantations where the Negroes are in a tremendous majority," she wrote in 1933, "the field hands have few contacts with white people and no need to amend their speech or give up the customs and traditions of their African ancestry. Living close to the earth, they know the ways of beasts and birds, clouds and winds, and interpret signs given them by the sun, moon and stars."[39]

Together these examples speak to *racialization,* the term used to describe the social processes through which ideas of race become reified and naturalized to explain and justify uneven power relations between people grouped together based on perceived differences in appearance. Race is not a biological "thing."[40] The idea of race emerged during New World colonization and was intimately tied to European imperialism and the enslavement of people from Africa on New World plantations. Still, up until the middle of the nineteenth century, slave status was more a legal distinction than one based on reified notions of race. For this reason, argued Grace Elizabeth Hale, "slavery founded and fixed the meaning of blackness more than any transparent and transhistorical meaning of black skin founded the category of slavery."[41] Enslaved African American laborers became racialized in a three-step process: (1) the classification of human populations based on perceived somatic differences, (2) the development of the idea that "natural" characteristics are innate and inheritable, and (3) the development of the idea that these "racial" differences were the *cause* of the unequal power relations between white masters and black slaves.[42] Through these examples it is suggested that postbellum social relations in the Lowcountry were similarly explained by appeal to the idea of superior and inferior races, the natural conditions of which supported, in the minds of low-country whites, the basic tenets of planter paternalism. If low-country African Americans were "naturally" childlike—Rutledge used the term "God's Children" to signify their innocence and need for paternal protection[43]—then low-country whites understood white supremacy as a simple expression of the natural order of things. This racist ideology also found expression in the landscapes of everyday life, where the visible presence of African Americans working in the fields and living in poverty may have confirmed for low-country whites the inferior status of the black race.

For Rutledge, as for other low-country whites, race explained everything. His stories extended nineteenth-century notions of race and tapped into early twentieth-century discourses on civilization and manliness of character. The discourse on civilization developed and maintained the idea that only a few races were civilized enough to rule or even to vote, a sentiment that formed the basis for Jim Crow laws across the South.[44] There was also an important gender dimension to the civilization narratives, which argued that advanced races were easily recognized by the degree of their sexual differentiation, particularly regarding the gendered spheres of (male) public space and (female) private space. Uncivilized men and women were considered indistinguishable from each other, either as likely to perform heavy labor in the fields. Again, the mere presence of African American women laboring on the postbellum plantations confirmed their racial inferiority in the eyes of low-country whites.

The gender ideology in the discourse on civilization belies an early twentieth-century American obsession with manliness expressed in the rapid growth of fraternal organizations; the growing popularity of camping, hunting, and fishing as middle-class pastimes; and the high sales of hunting and adventure stories as well as "success" manuals detailing the cultivation of "manliness of character."[45] Many of Rutledge's stories appeared in fishing, hunting, and outdoor magazines and blended the discourse on manliness with that of planter paternalism.[46] Manly traits such as assertiveness, decisiveness, determination, and self-motivation were tempered with wisdom, generosity, and protective concern for everyone on the plantation.

Although African American women are absent from Rutledge's hunting stories, they appear in many of his plantation tales as hardworking field hands. According to Rutledge, these women work mostly as a result of the African American male's failure to fulfill his manly duty as family provider. Although the men on Hampton were adept at the hunt, Rutledge believed they were naturally lazy, inheriting a disdain for work from generations of "meek servitors." Giving advice to "thousands of other plantation owners," Rutledge wrote that low-country African American men "always feel that work is a kind of imposition, obtruding itself on the joy of life; it is always secondary to the grand primary business of getting happiness out of life."[47]

In defining white men as shooting participants of the hunt, as heads of households, as patriarchs, and as providers and protectors of their black and white family, Rutledge reveals much about low-country men as gendered and racialized beings. His work also illustrates the extent to which art, narratives, and the presentation of heritage landscapes maintain gendered and racialized social relations. Through his hunting and plantation stories, Rutledge not only defined the racial and gender categories of the people living on low-country plantations but taught a way of seeing the Lowcountry, where these categories were made to seem natural, as part of the visible landscape. The low-country plantation landscape reflected then, and continues to reflect in heritage landscapes today, these taken-for-granted ideas about gender and race. Rutledge's stories advanced a particular way of seeing the low-country landscape that naturalized white male power. The Hampton Plantation landscape, then, is the planter discourse materialized.[48]

Alternative Landscape Narratives

Adam Goodheart wrote that plantation heritage landscapes possess equal measures of nostalgia and amnesia.[49] Only until very recently, for example, the Rice Museum in Georgetown, South Carolina, never mentioned the word *slave* in its exhibits on low-country rice culture, preferring the euphemism *agricultural laborer* to describe the enslaved laborers depicted in the plantation dioramas. Such forgetting is accompanied by a strong sense of nostalgia, produced through white reverence for an imagined Old South myth that is in turn dependent on a careful erasure of the brutal social relations that produced southern white wealth. This is not surprising, given that many plantation heritage sites were established through the efforts of white women, many of whom traced their ancestry back to the slaveholding families associated with the plantations they helped preserve.[50]

Physical landscape artifacts at Hampton that are missing from the visible scene—the rice fields, the massive flood control structures (trunks), the slave settlement and street—all present challenges to overcoming nostalgia and amnesia on this plantation heritage site. Recent scholarship has established the crucial role played by enslaved African Americans in shaping this landscape, not only through their coerced labor but also from their knowledge of rice cultivation developed in West African rice-growing regions.[51] With some effort, this landscape could be made to tell a parallel story to that of Rutledge's life, although this narrative includes many less than pleasant images of our shared history.

Rutledge's stories, like Smith's watercolors, continue to present seductive messages of white superiority in the Lowcountry. This is due in large measure to the beauty of the art that delivers the message. Few who read Rutledge's stories will doubt his sincere belief in the benevolence of the plantation myth or will doubt his affection for those African Americans living on Hampton, particularly members of the Alston family. Likewise, those who see Smith's watercolor paintings will no doubt appreciate the skill with which she depicts the beautiful plantation scenes. But the voices of African Americans are silenced in these picturesque plantation landscapes or are filtered through white sensibilities. Artistic success in painting or in prose seems built on an aesthetic that ignores the troubled historical geography of the low-country landscape. Goodheart further observed that this erasure emerges from a paradox: "Some of this nation's cruelest places can also be its most beautiful."[52] This purposeful forgetting presents a benign, even sanitized, plantation landscape for mostly white tourist consumption.

The physical erasure at Hampton Plantation is accompanied by a narrative elision. Focusing on the life of Archibald Rutledge and the architectural history of the big house, to the exclusion of other possible stories, continues to define whiteness in the terms of the Jim Crow era.[53] The material landscape at Hampton Plantation cannot be expected to communicate this complex history without developing educational narratives that teach visitors a different way of seeing this landscape. One encouraging sign is the development of teaching materials for South Carolina schoolchildren asking that they imagine life as an enslaved laborer on the rice plantation. The meaning of any landscape never inheres solely in the material objects that together comprise the visible scene[54] but is constructed through the stories told about the cultural landscape.

It is difficult to imagine plantation heritage sites directly confronting the brutal history of slavery, the construction of whiteness, and the maintenance of white supremacy, although some have begun to incorporate slave narratives into their landscape interpretations. Brookgreen Gardens, a private plantation heritage site and sculpture garden just north of Georgetown, South Carolina, has reconstructed an old plantation rice field, complete with functioning rice trunks and gates. Visitors to the boardwalk overlooking the rice field are surprised with a recorded message from a fictional slave Ben, who explains the technical expertise needed to mind the trunks that regulate the flooding and draining of the rice fields and the special status this affords the enslaved "trunk-minder." In the recorded message, Ben also relates his fears that the plantation master will sell his son off-plantation unless he too can learn to mind the trunks. Brookgreen is best known as a formal garden that displays the work of early twentieth-century representational sculptors, and the beautiful parklike landscape does not prepare the visitor for the poignant monologue. Presenting this narrative apart from the main landscape circulation, however, signals what Eichstedt and Small called "the rhetorical strategy of segregation," where the "black" history told in this one corner of the plantation landscape is not allowed to permeate the rest of the site.[55] Slavery is presented as an evil but somewhat faceless institution exerting abstract power over the lives of enslaved families. And yet, as Elizabeth Fox-Genovese pointed out, "that power was no abstraction: It wore a white, male face."[56]

Plantation heritage sites contribute to a racialized representational regime. At Hampton, narratives about the production of the material landscape are erased, together with the experiences of the many African Americans who called Hampton home. What remains are framing devices that valorize the experience of whites. Rutledge's life and his stories of Hampton communicate a partial landscape history, one embedded in white cultural memory. Perhaps the landscape at Hampton Plantation could support an interpretation of the history of racism, exposing its underlying ideologies and the many ways the plantation myth—as a mixture of discourses on race, gender, civilization, and paternalism—has been, and continues to be, instantiated in the material landscape.

The low-country rice plantation has always been both an economic project and a racial project. The slow postbellum decline in rice production, however, diminished its economic importance. Interest in plantations as heritage landscapes, however, remains high—as evidenced by the nearly continuous publication of coffee-table books featuring low-country rice plantations.[57] Even more important, the low-country plantation's status as one important driver of the coastal tourism industry elevates the symbolic importance of sites such as Hampton Plantation. The African American majority that produced and gave specific cultural meanings to these landscapes, however, continues to be excluded from plantation heritage narratives. Without dramatic changes in the way these public sites are presented, low-country rice plantations will remain racial projects that materialize the ideology of white supremacy, through the discourse of planter paternalism, in the visible material landscape.

Notes

1. Jim Casada, ed., *Hunting and Home in the Southern Heartland: The Best of Archibald Rutledge* (Columbia: University of South Carolina Press, 1992); and Archibald Rutledge, *Tom and I on the Old Plantation* (New York: Frederick Stokes, 1918).
2. Elizabeth Allston Pringle, *Chronicles of Chicora Wood* (Boston: Christopher, 1940), 67.
3. Charles Joyner, *Down by the Riverside: A South Carolina Slave Community* (Urbana: University of Illinois Press, 1984).
4. Georgia Writer's Project, *Drums and Shadows: Survival Stories among the Georgia Coastal Negroes* (Athens: University of Georgia Press, 1940).
5. Julia Peterkin, *Roll, Jordan, Roll* (New York: Ballou, 1933), 9.
6. Archibald Rutledge, *Home by the River* (Bobbs-Merrill, 1941; repr., Orangeburg, SC: Sandlapper, 1983), 13.
7. For more about the different ways of seeing the low-country landscape, see Samuel F. Dennis Jr., "Seeing the Lowcountry Landscape: 'Race,' Gender and Nature in Lowcountry South Carolina and Georgia, 1750–2000" (diss. in geography, Pennsylvania State University, 2000).
8. Julie Saville, *The Work of Reconstruction: From Slave to Wage Laborer in South Carolina, 1860–1870* (Cambridge: Cambridge University Press, 1996).
9. Betty Wood, *Women's Work, Men's Work: The Informal Slave Economies of Lowcountry Georgia* (Athens: University of Georgia Press, 1995).
10. Patricia Guthrie, *Catching Sense: African American Communities on a South Carolina Sea Island* (Westport: Greenwood, 1996).
11. Denis Cosgrove, "Prospect, Perspective and the Evolution of the Landscape Idea," *Transactions of the Institute of British Geographers* 10 (1985): 45–62.
12. John Michael Vlach, *The Planter's Prospect: Privilege and Slavery in Plantation Paintings* (Chapel Hill: University of North Carolina Press, 2002); and R.H. Schein, "Representing Urban America: 19th-Century Views of Landscape, Space, and Power," *Environment and Planning D: Society and Space* 11 (1993): 7–21.
13. Rutledge, *Home by the River*, 15.
14. Robert Mills, *Mills' Atlas of the State of South Carolina, 1825* (Easley, SC: Southern Historical Press, 1980).
15. Dell Upton, "White and Black Landscapes in Eighteenth-Century Virginia," *Places* 2 (1985): 59–72.

16. John Michael Vlach, *Back of the Big House: The Architecture of Plantation Slavery* (Chapel Hill: University of North Carolina Press, 1993).
17. Rutledge, *Home by the River.*
18. Sam B. Hilliard, "The Tidewater Rice Plantation: An Ingeneous Adaptation to Nature," *Geoscience and Man* 12 (1975): 57–66; Daniel C. Littlefield, *Rice and the Making of South Carolina: An Introductory Essay* (Columbia: South Carolina Department of Archives and History, 1995); and Judith Carney, *Black Rice: The African Origins of Rice Cultivation in the Americas* (Cambridge, MA: Harvard University Press, 2001).
19. Jennifer L. Eichstedt and Stephen Small, *Representations of Slavery: Race and Ideology in Southern Plantation Museums* (Washington, D.C.: Smithsonian Institution Press, 2002).
20. Casada, *Hunting and Home in the Southern Heartland;* and Dorothy Stone Harmon, *Archibald Rutledge: The Man and His Books* (Piedmont, SC: Bookquest, 2003).
21. Rutledge, *Home by the River,* 13.
22. Michael Foley, Marion Edmonds, and Ray Sigmon, *Hampton Plantation State Park Visitors Guide* (Columbia: South Carolina Department of Parks, Recreation, and Tourism, 1983).
23. Rutledge, *Home by the River,* 27.
24. Foley, Edmonds, and Sigmon, *Hampton Plantation State Park Visitors Guide.*
25. Rutledge, *Home by the River,* 162.
26. George C. Rogers Jr., *The History of Georgetown County, South Carolina* (Columbia: University of South Carolina Press, 1970).
27. Pringle, *Chronicles of Chicora Wood,* 17.
28. Raymond Williams, *The Country and the City* (Oxford: Oxford University Press, 1973).
29. Rutledge, *Home by the River,* 164–65.
30. Ibid., 13.
31. Wood, *Women's Work, Men's Work,* 127.
32. Alice Ravenel Huger Smith, quoted in Martha R. Severens, *Alice Ravenel Huger Smith: An Artist, a Place and a Time* (Charleston, SC: Carolina Art Association, 1993), 97.
33. Herbert Ravenel Sass, *A Carolina Rice Plantation of the Fifties* (New York: William Morrow, 1936).
34. George P. Rawick, ed., *The American Slave: A Composite Autobiography* (Westport, CT: Greenwood, 1972).
35. Peterkin, *Roll, Jordan, Roll,* 20.
36. Rutledge, *Home by the River,* 37.
37. Ibid., 47.
38. Ibid., 47–48.
39. Peterkin, *Roll, Jordan, Roll,* 23.
40. Michael Banton, *The Idea of Race* (London: Travistock, 1977); Colette Guillaumin, "The Idea of Race and Its Elevation to Autonomous Scientific and Legal Status," in *Sociological Theories: Race and Colonialism* (Paris: UNESCO, 1980); Stephan Jay Gould, *The Mismeasure of Man* (1981; revised

and expanded, New York: Norton, 1996); Robert Miles, *Racism and Migrant Labor* (London: Routledge, 1982); Peter Jackson, "The Idea of 'Race' and the Geography of Racism," in *Race and Racism: Essays in Social Geography*, ed. Peter Jackson (London: Allen and Unwin, 1987); Peter Jackson and Jan Penrose, eds., *Constructions of Race, Place and Nation* (Minneapolis: University of Minnesota Press, 1993); Robert Miles, *Racism after "Race Relations"* (London: Routledge, 1993); and David Brion Davis, "Constructing Race: A Reflection," *William and Mary Quarterly*, 3rd series, LIV (1997): 7–18.

41. Grace Elizabeth Hale, *Making Whiteness: The Culture of Segregation in the South, 1890–1940* (New York: Pantheon Books, 1998), 4.
42. See Guillaumin, "The Idea of Race and Its Elevation to Autonomous Scientific and Legal Status."
43. Archibald Rutledge, *God's Children: My Negro Friends at Hampton* (New York: Bobbs-Merrill, 1947).
44. Gail Bederman, *Manliness and Civilization: A Cultural History of Gender and Race in the United States, 1880–1917* (Chicago: University of Chicago Press, 1995).
45. J.A. Mangan and James Walvin, *Manliness and Morality: Middle-Class Masculinity in Britain and America, 1800–1940* (New York: St. Martin's, 1987); Mark C. Carnes and Clyde Griffin, eds., *Meanings for Manhood: Constructions of Masculinity in Victorian America* (Chicago: University of Chicago Press, 1990); Anthony Rotundo, *American Manhood: Transformations in Masculinity from the Revolution to the Modern Era* (New York: Basic Books, 1993); and Judy Hilkey, *Character Is Capital: Success Manuals and Manhood in the Gilded Age* (Chapel Hill: University of North Carolina Press, 1997).
46. See Archibald Rutledge, *Hunter's Choice* (New York: A.S. Barnes, 1946); and Casada, *Hunting and Home in the Southern Heartland.*
47. Rutledge, *Home by the River*, 111.
48. R.H. Schein, "The Place of Landscape: A Conceptual Framework for Interpreting an American Scene," *Annals of the Association of American Geographers* 87 (1997): 660–80.
49. Adam Goodheart, "The Bonds of History," *Preservation* 53 (September/October 2001): 36–43, 94.
50. Eichstedt and Small, *Representations of Slavery.*
51. Judith Carney and Richard Porcher, "Geographies of the Past: Rice, Slaves, and Technological Transfer in South Carolina," *Southeastern Geographer* 33 (1993): 127–47; and Judith Carney, *Black Rice.*
52. Goodheart, "The Bonds of History," 37.
53. Steven Hoelscher, "Making Place, Making Race: Performances of Whiteness in the Jim Crow South," *Annals of the Association of American Geographers* 93 (2003): 657–86.
54. Trevor Barnes and James Duncan, "Introduction: Writing Worlds," in *Writing Worlds: Discourse, Text, and Metaphor in the Representation of Landscape*, ed. Trevor Barnes and James Duncan (London: Routledge, 1992).
55. Eichstedt and Small, *Representations of Slavery.*

56. Elizabeth Fox-Genovese, *Within the Plantation Household: Black and White Women of the Old South* (Chapel Hill: University of North Carolina Press, 1988), 190.
57. Samuel Galliard Stoney, *Plantations of the Carolina Low Country* (1938; rev. ed. Charleston: Carolina Art Association, 1955); Alberta Morel Lachicotte, *Georgetown Rice Plantations* (1955; 7th printing, with revisions, Georgetown, SC: Georgetown County Historical Society, 1993); Suzanne Cameron Linder and Marta Leslie Thacker, *Historical Atlas of the Rice Plantations of Georgetown County and the Santee River* (Columbia: South Carolina Department of Archives and History, 1995); and William P. Baldwin, *Lowcountry Plantations Today* (Greensboro, NC: Legacy, 2002).

CHAPTER 5

Poetic Landscapes of Exclusion: Chinese Immigration at Angel Island, San Francisco

GARETH HOSKINS

Suspended on a wooden frame at the edge of a small beach on Angel Island in the San Francisco Bay is a two-ton lump of oxidized gray-green bronze known as the Immigration Station Bell. It hangs as a pivot between two long-destroyed but constantly invoked features: a 120-foot wharf reaching out into the bay and, on land, a two-story federal administration building.[1] Today the bell's clapper is swung occasionally by visitors, each time adding a dent to the thousands of pockmarked scars from years of barrage. The chime produced is a rewarding clear tone that carries out across the water to the mainland, a sound that originally acted as a warning signal for traffic negotiating the narrow fog-plagued Raccoon Straight, a passage that joins the Sacramento River with the Pacific Ocean. But now, as well as being a physical prop for the aural pleasure of tourists to a heritage site, the bell is a gathering point for the beginning of a tour that brings into collision ideas of race, national identity, and exclusion. From the Immigration Station Bell, visitors are guided by California State Park docents on a journey tracing the footsteps of an estimated 175,000 Chinese immigrants that were detained on the island between 1910 and 1940; immigrants processed under racially targeted laws that severely restricted their entry to the United States and prevented them from obtaining citizenship. Typically, this tour begins with the following words:

> Imagine a typical foggy day in the San Francisco Bay. The bell rings automatically to warn passing ships of the wharf. You have travelled for about twenty-one days across the Pacific, leaving your family and all you know behind. The immigration station's ferryboat, the Angel Island arrives with you and fifty others on board. You get off the boat and onto the wharf where you see a huge building with two large flagpoles out front. You see a fifteen-acre site completely fenced in. You walk to the front porch of the administration building. This building will begin the start of your enclosed life.[2]

Imagining the bell's tone as it echoes around the shores of a San Francisco Bay in the 1910s provides a powerful dual point of departure for the listener. A temporal shift from the present to the past is accompanied by a corporeal shift where the visitor is asked to *become* the immigrant, to adopt her perspective, occupy his bodily space, and experience their sensations. "You have traveled for about twenty-one days," "you get off the boat," "you walk to the front porch"—this is "your" life of enclosure.

Visitors certainly have to use their imagination. Laid out in front of them is an unremarkable expanse of cracked concrete foundations joined by overgrown paths with deteriorating military buildings visible through the trees. From here the only hint of this place's significance is an ambiguous black granite monument rising eight feet from the center of some open ground. However visually benign, the Angel Island Immigration Station continues to play an important role in negotiations about who is allowed to be considered an American and the manner in which they are considered as such.

The Angel Island Immigration Station is a racialized landscape in a double sense. First, from 1910 to 1940 it acted as a materialized expression of exclusion. It worked to defend the nation's borders from a perceived Asian menace and, by association, served to construct the legitimate American public as white. Second, as a historic site today, the Angel Island Immigration Station landscape is put to work telling an uncomfortable chapter in the nation's immigrant history and provoking debates on citizenship, race relations, and contemporary immigration policy. In each case, it is a landscape where racial categories frame a discourse of national identity.

That this landscape exists to inform debates about race and identity in the United States at all is due, in part, to one moment of serendipity. In 1970 Alexander Weiss, then a ranger at Angel Island State Park, was in the process of securing a dilapidated, two-story clapboard barrack listed in the park's inventory as "Building 317" and earmarked for demolition. Moving through the barrack, the oblique angle of Weiss's flashlight picked up indentations of what appeared to be Chinese letters all along its walls. Snubbed when he informed his superiors of the possible significance his finds, Weiss called on his college professor, Dr. George Araki, a Japanese American, to view the inscriptions. Soon after, Araki and photographer Mak Takahashi began the process of recording what turned out to be hundreds of poems carved by Chinese immigrants detained by the federal government under suspicion of trying to evade the 1882 Chinese Exclusion Act.

This was not the first time the poems were brought into the public domain. More than thirty years earlier when the building still was an immigration station, two Chinese immigrants, Smiley Jan detained in 1931 and Tet Yee detained in 1932, copied more than ninety poems in an unpublished manuscript titled "Collection of Autumn Grass: Volume Collecting Voices from the Hearts of the Weak."[3] Selections from this version of the poems found their way into the journals *Asian American Review, Frontiers,* and *Amerasia.*[4] But it was Weiss's detective skills and his chance moment of rediscovery that emboldened members of the local Chinese American community to form the Angel Island Immigration Station Historical Advisory Committee (AIISHAC), which within two years had secured $250,000 from the state through grant measure AB3067 for the building's protection and stabilization.

Today, these accounts of immigrants' experiences carved into the barrack walls have become part of the national heritage landscape, deemed to be significant national treasures. They capture firsthand an important moment in Chinese American history and also provide exceptional material evidence with which to communicate how racial exclusion played a role in building the United States.

The poems are largely the product of male immigrants who journeyed to America's Pacific shore from the Cantonese villages of the Pearl River Delta region in Guangdong Province, South China. They follow the conventions of the classical Tang dynasty style, with four or five stanzas employing sophisticated allegorical references that evoke characters in Chinese myth and legend to represent the frustration, hostility, and pain of being imprisoned (see Figure 5.1). One particularly eloquent poem carved at head height in a former barrack bathroom is translated on an interpretive board. It reads,

> Pity it is that a hero has no way of exercising his power.
> He can only wait for the word to whip his horse on a homeward journey.
> From this moment on, we say goodbye to this house.

Figure 5.1 The most visible of all the poems in the barrack building. Photograph by the author.

My fellow countrymen here are rejoicing like me.
Say not that here everything is western styled.
Even if it were built with jade, it has turned into a cage.[5]

Another poem found in the men's dormitory tells us,

This place is called the island of immortals,
When, in fact, this mountain wilderness is a prison.
Once you see the open net, why throw yourself in?
It is only because of empty pockets I can do nothing else.[6]

Others present more extensive biographical accounts:

A member of the Li household was ready to leave.
In the last month of summer, I arrived in America on ship.
After crossing the ocean, the ship docked and I waited to go on shore.
Because of the records, the innocent was imprisoned in a wooden building.
Reflecting on the event, my heart is vexed and depressed.
I composed a poem to rid myself of sadness and worry.
At present, my application for admission has not yet been dismissed.
As I record the cause of my situation, it really provokes my anger.
Sitting here, uselessly delayed for long years and months,
I am like a pigeon in a cage.[7]

Although it is tempting to understand these verses as expressions of lament with the author as passive victim forced into silence, many poems express vehement anger toward the United States.

> When a newcomer arrives in America
> He will surely be seized and put in the wooden building,
> Like a major criminal,
> I have already been here one autumn.
> The Americans refused me admission;
> I have been barred and deported back.
> Alongside the ship, the waves are huge.
> Returning to the motherland is truly distressing.
> We Chinese of a weak nation,
> Sigh bitterly at the lack of freedom.
> The day our nation becomes strong,
> I swear I will cut off the barbarians' heads.[8]

Indeed, these poems were seen at the time as a form of active resistance, as evidenced in a letter by Frank Hays, head of the Deportation and Detention Division, who wrote to the commissioner of immigration at the immigration station:

> I notice the walls in the rooms of the general quarters have been considerably marred by the aliens writing on them and the property has otherwise been disfigured and destroyed. It is respectfully suggested that appropriate signs or notices be placed in different rooms worded in such a manner that the aliens will understand the impropriety of injuring the property in any way. The following sample is submitted.
>
> Notice:
>
> This building belongs to the United States Government. It is unlawful to write on or disfigure the walls or to destroy any property on these premises.[9]

Today the poems form the climax of a tour that takes the visitor around the immigration station remains and through the barrack rooms where the immigrants were held. In this context, the poems establish the landscape as a material expression of Chinese exclusion—a place where the Chinese were pathologized, constructed as a threat, to the national community. Here the landscape reveals a legacy of intolerance and oppression sanctioned at the highest level of government.

Conversely, that the Angel Island Immigration Station functions today as one of the more progressive National Historic Landmarks in America's formal heritage landscape indicates huge possibilities for empowerment. Connecting the Chinese American community directly to the production of their historical landscapes can help to democratize remembrance and in the process realign national memory to include a particular immigration story that challenges a universalized celebratory myth of immigrant America. Communicating the contradictory experience of Chinese immigrants as part of national immigration history can invigorate public debate about race and nation where race is understood not as an ontologically pregiven category based on biological essence but as a socially constructed outcome of historically contingent modes of oppression. Being conscious of the mutually constitutive relationship between race and space and thinking through those relationships to understand how the landscape is implicated in race-making events is crucial because it provokes questions about how immigrant processing, for example, both renders whiteness invisible and racializes others. Hence, by tracing the production of the Angel Island Immigration Station landscape, we can achieve a better understanding of the power relations that work to sustain race as a viable category. Such efforts entail conceiving of the landscape as a landscape of power, a material articulation of the construction(s) of race. In the next section I undertake a landscape history of the Angel Island Immigration Station to show how Chinese immigrants have been implicated into a politics of exclusion.

Creating Angel Island

Angel Island is the largest island in the San Francisco Bay. It has 740 acres, has six miles of shoreline, and looks through the Golden Gate, rising steeply from sea level to 788 feet at its peak, Mount Livermore (see Figure 5.2). Today, the island is a popular destination for visitors; 200,000 of them arrive each year on public ferries and private boats to enjoy the miles of trails and sheltered picnic grounds that give access to a varied natural history as well as impressive views of Alcatraz, Marin County, the East Bay, and the San Francisco skyline. The most frequently explored part of the island is the U.S. Immigration Station, accommodating 27,000 people annually.[10] The majority of these visitors are elementary school children on scheduled tours. On weekends, however, many elderly Chinese Americans, encouraged by their relatives, return to look around the remains of a place that holds powerful personal memories.

Although today the immigration station and the Chinese immigrant experience in particular are fast becoming synonymous with the whole of

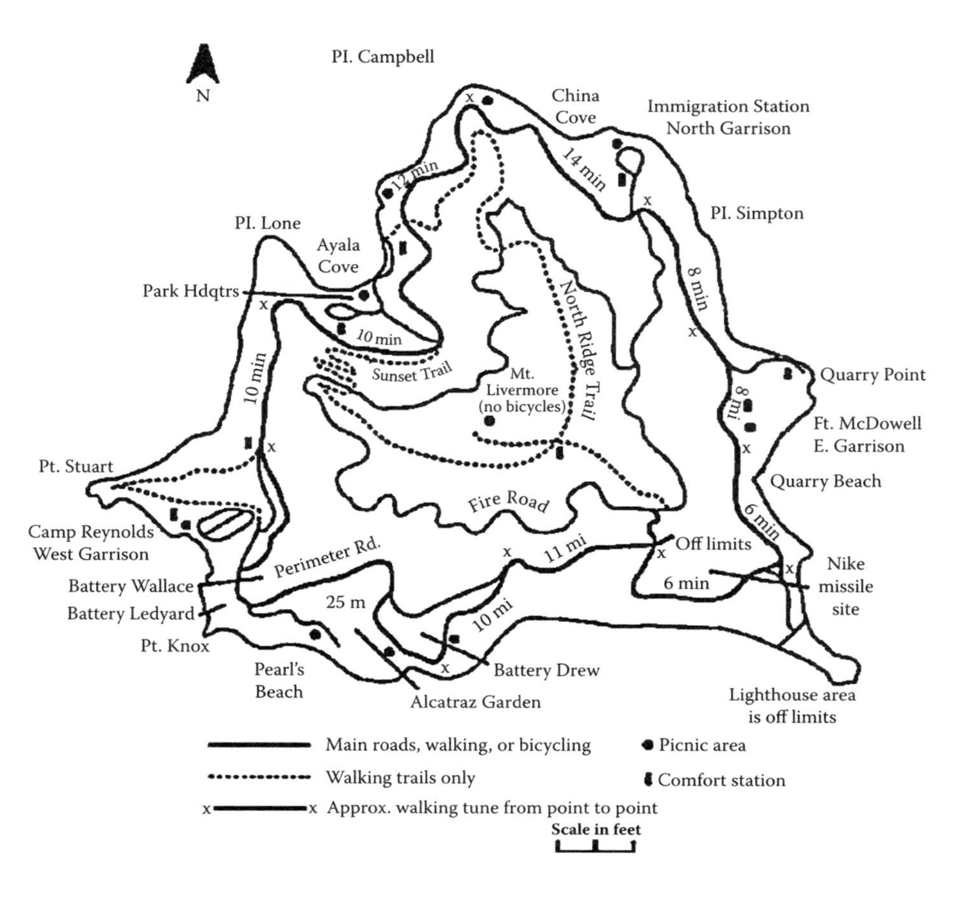

Figure 5.2 Angel Island. Source: California Department of Parks and Recreation.

the island, the landscape has a long and varied human history. Evidence of seasonal hunting and fishing by the Hookooeko Tribe of the Coast Miwok Indians suggests that the island was populated at least one thousand years ago. The island was placed in the historical record in 1775 when Manuel de Ayala, a lieutenant in the Spanish navy, used it as a base from which to survey the San Francisco Bay. As in so many other cases, the island's "discovery" by Europeans opened up the area for Spanish settlers, bringing with them diseases that decimated the resident Miwok population. Since that time, different groups have taken advantage of the island's useful location in various ways: Russian sea-otter hunters worked on the island, European settlers occupied the land for ranching, and a Chinese fishing community was located there. By 1850 the island had become more valuable for its strategic position when President Millard Fillmore declared it a military reserve. This began a series of occupations by the armed forces that lasted until the Cold War. The island became home to two army garrisons: Camp Reynolds from 1863 was established on the western shoreline out of concern over threats to the Bay Area by Confederate sympathizers, and Fort McDowell, looking to the east, was established in 1899 as a detention center for captives of the Spanish American War. The potential for Angel Island as a secure immigrant processing location was recognized by the early twentieth century, spurred by the success of New York's Ellis Island. During the period of its operation as an immigration station (1910–40), Angel Island processed groups from all corners of the globe, including Japanese picture-brides and immigrants from India, Korea, the Philippines, and Europe, but it was the Chinese arriving there in far greater numbers who were singled out for special scrutiny.

In the late nineteenth century, Chinese immigrants working in California were made scapegoats for successive economic depressions. The 1882 Chinese Exclusion Act signed by President Chester Arthur was a response to the demands of a militant white labor force reacting to the high unemployment and low wages sparked by post-Civil War depression and an increasingly competitive labor market brought by so many new arrivals to the West. The Exclusion Act was the centerpiece of a series of racist legislation designed to prohibit the entry of Chinese laborers into the country for ten years and to deny citizenship rights to all Chinese already in the United States. The act was extended periodically until 1904, when it was installed indefinitely. In 1888 the Scott Act prohibited any reentry of Chinese laborers, and in 1892 the Geary Act decreed that all Chinese caught illegally residing in the United States would be deported following one year of hard labor. The Exclusion Act was not repealed until 1943, through China's cooperation with the United States in the Second World War. Even

Figure 5.3 The administration building circa 1925. Source: Immigration station lantern slides, California Department of Parks and Recreation.

then, individuals of Chinese descent often were detained to determine the validity of their entry papers.

The 1882 Chinese Exclusion Act was the first functional U.S. national policy on immigration, and it meant that operational responsibility would be shifted to the federal level, thus instituting a far more formal system of processing than that previously offered by California state officials and the Pacific Mail Steamship Company at dockside sheds on San Francisco's wharfs.

The Bureau of Immigration selected Angel Island as a site for the new U.S. immigration station in 1904, but the earthquake of 1906 and repeated administrative problems delayed its opening until January 21, 1910. When completed in 1914 the immigration facility consisted of a 120-foot wharf with a baggage house, a large administration building, a detention barrack, a powerhouse, a hospital, and cottages for the facility's employees (see Figure 5.3). The race-based inspection procedures mandated under the Exclusion Act meant that all Chinese travelers whatever their social status were regarded with suspicion by the authorities. A ship arriving in San Francisco containing Chinese passengers, for instance, would be boarded by an immigration official who collected documents and questioned each individual seeking to land. Those failing to give satisfactory proof of their right to enter were sent to Angel Island for further scrutiny. Once at the island, immigrants boarded the pier and were relieved of their baggage, which was then searched. The facility operated strict separation procedures where Chinese immigrants were kept apart from Japanese and other Asians and males were separated from females until their cases were concluded to prevent collusion between spouses.

Immigrants began their processing with medical examinations conducted by the U.S. Public Health Service, which looked for visible indicators of conditions such as hookworm, roundworm, trachoma, and liver flukes that would provide grounds for exclusion. In this way medical officers were responding to and participating in the conflation of immigrant with disease, the widely held belief that the Chinese were a greater health risk than other racial groups—a perspective evident in sporadic health scares and fumigations of Chinatown in San Francisco.[11] Following the medical examination Chinese immigrants were moved to the detention barrack, where they would await a legal hearing in front of an immigration inspector, prison guard, translator, and stenographer. Because the Exclusion Act was directed against Chinese laborers specifically, an immigrant's case rested on providing evidence that he or she was a natural-born American citizen, a child of a citizen, or a member of a class exempt from the exclusion law, a list that included merchant, student, teacher, diplomat, and traveler. In practice, the policy was such that the burden of proof fell on the immigrants at arrival to provide evidence that they were not excludable. Convincing skeptical officials charged with enforcing the laws was difficult. A case might be heard a number of times where answers provided to an inspector were compared to answers given by witnesses vouching for the applicant's identity. Carlton Rickards, an official dealing with Chinese immigration through the port of San Francisco, testified to a congressional committee in 1890 about the procedures he employed for inspection: "Of course I try to get evidence not for the Chinaman but against him, and then he has got to make his own proof. … My examination is taken for the collector [of customs] and is as much against the case as it can be."[12]

Although the exclusion laws drastically reduced the number of Chinese applying for entry into the United States, most of those who did apply were eventually admitted. Their experience at Angel Island, however, left them with little doubt about the national antipathy to their presence. Oral history undertaken to illustrate Him Mark Lai's collection of poems *Island* (1991) provides personal accounts of the detention experience and gives us insight into the feelings of immigrants enduring this treatment.

> I had nothing to do there. During the day, we stared at the scenery beyond the barbed wires—the sea and the sky and clouds that were separated from us. Besides listening to the birds outside the fence, we could listen to records and talk to old timers in the barracks. Some, due to faulty responses during the interrogation and lengthy appeal procedures, had been there for years. They poured out their

> sorrow unceasingly. Their greatest misery stemmed from the fact that most of them had to borrow money for their trips to America. Some mortgaged their houses; some sold their land; some had to borrow at such high interest rates that their family had to sacrifice. A few committed suicide in the detention barracks. The worst part was the toilet. It was a ditch congested in filth. It stank up the whole barracks. We slept there in three tiers of canvas bunks. The blankets were so coarse that it might have been woven from wolf's hair. It was indeed a most humiliating imprisonment.[13]

Comments about the poor conditions of the Angel Island Immigration Station were not limited to the detainees. Although Angel Island was intended as a state-of-the-art and up-to-date processing facility, even health inspectors noted its appalling sanitary conditions, citing inadequate structural design, bad maintenance, a shortage of fresh water, and chronic overcrowding. After one external review in 1922, the commissioner general of immigration commented, "Angel Island is the worst Immigration Station I have visited … the sanitary arrangements are awful. If a private individual owned such an establishment he would be arrested by the local health authorities."[14] Problems beset the immigration station continually; all maintenance and operating costs were doubled because of the awkwardness of transportation, and the Chinatown community repeatedly complained that the island was an unfair and inconvenient location for the processing. So in 1940 when the main administration building of the immigration station was destroyed by fire, there was little will to continue with immigrant processing on Angel Island, and the immigrants once again went back to detention facilities in the city, first to Silver Avenue and then to Sansome Street. With the administration building leveled, the military once again assumed control over the site in 1941, making use of the remaining buildings by stationing troops on their way to war and securing German, Japanese, and Italian prisoners.

Landscape as Heritage

At the end of the war, the island became redundant as a military garrison and was abandoned until 1948 when the Department of the Interior assumed responsibility for the island, allowing it to be turned over to the public as a historical monument. Two years later in 1950, the Angel Island Foundation was formed, which began a campaign to turn the island into a state park. California parks were few at the time, and although Angel Island was a valuable addition to the state's portfolio of natural and aesthetic resources, the administration took on responsibility for the island

Figure 5.4 The barrack in 2001. Photograph by the author.

somewhat reluctantly and in portions. Under an operation remit of recreational tourism, the park closed for a three-month period in 1957, during which time many buildings were demolished and their debris burned and buried. In 1963 the state of California received title to 517.24 acres of Angel Island, allowing another 110 buildings to be razed and burned. Fiscal limitations of the park system at the time required that the island provide some sort of economic return, and with this in mind eucalyptus stands were planted for timber that by 1970 began to dominate the former immigration station site.

The rediscovery of the Chinese poems in 1970 established the island's relevance to the Chinese American community as a site of memory. In 1983 two part-time rangers worked to open the barrack building for informal public tours (see Figure 5.4). A San Francisco wax museum donated mannequins, Chinese-style objects were displayed to help bring alive the barrack's sparse interior, and a forty-page document was put together to assist a growing number of guides offering tours.[15]

By the end of the 1970s, research by Him Mark Lai, Genny Lim, Judy Yung, George Araki, and others provided a great deal of information about the poems and the experiences of the immigrants detained on the island. At the same time, however, a sense of shame and fear of reprisal felt by some previous detainees demanded that the story be told with sensitivity. Introductory comments in *Island* (1991) concede,

> As a whole, the former detainees hesitated to reveal an unpleasant past they preferred left forgotten. It was only after a promise of anonymity that they agreed to be interviewed for this book.[16]

Paul Chow, a leading figure in the early years of the restoration project, told reporter Katherine Bishop in an interview for the *New York Times* that "the generation that went through [the immigration station] wants to forget it. It's the first and second generation that says this was shameful and you put my parents and grandparents through this and it should not be forgotten."[17]

For ten years Angel Island advocates consolidated their local profile, holding various commemorative events and reunions for former detainees until an ambitious reassessment of the organization's role began that would transform how the immigration station's history and the exclusion experienced there would be remembered. In 1996 Daniel Quan, then on the board of directors of the Angel Island Immigration Station Foundation, a contemporary offshoot of the AIISHAC, spoke to the local press about changes occurring at the time:

> For a long time, [Chow] was all alone in this project and it's hard to be a one man crusade. Some of the people involved today are in sync with [Chow's] vision, but we have our sights on making this a widespread organisation that reaches more people around the country.[18]

In October 2001 I asked Quan to reflect on that time of transition, and he spoke with me frankly about his feelings:

> We had tried small for a number of years and small wasn't working for me. I was thinking "well we're not really making a big dent and we're doing the same thing year after year and so what's the harm in going big." ... Previous to that the board had been stacked in the other direction. All the members were passive and wanted to keep everything with the status quo, pretty much exactly how it had been operating for a number of years. Personally it was starting to feel a little meaningless, like you were existing just to exist as an entity so that you could call yourself a non-profit but you weren't actually bringing about any change.[19]

With Quan at the helm, the foundation became proactive, transforming its relationship with the site and embarking on successive campaigns to springboard the immigration station into the national arena. The main agenda was to repackage the landscape in ways that would reframe the scale at which the history of Chinese exclusion was identified. The campaign would take a particular direction.

> I saw establishment of the [National] Landmark Status as being the key to the success for grants and anything else that would

> happen afterwards. We would have to elevate it to the national status in order to get any kind of recognition or grant monies at all.[20]

Another member of the foundation's board shared the same vision:

> Elevating the credibility of this history with the park service allowed us to swim in other arenas. So there was some strategic thinking about how we start stepping up to the plate. And how we participate in those waters that we are now allowed to enter.[21]

Achieving National Historic Landmark status for the immigration station required making a case to the National Registry in a way that would transform how the site was understood. What was, for many, still a place of personal and private trauma would now be presented to the public as a nationally significant resource where healing, learning, and self-affirming cries of injustice would extend to all peoples of the United States oppressed on account of their race.

In 1997 the immigration station became a National Historic Landmark, providing a new level of prestige and inserting the landscape into the national arena, thus making it eligible for a host of new funding initiatives. From that moment, the immigration station's public profile increased and momentum built for its restoration. In the year 2000 California state senators inserted a line item into a $2.1 billion bond measure that was approved by public referendum, guaranteeing the immigration station $15 million dollars in funds for planning and restoration work. Although these funds are only half of the amount needed to complete what have become ambitious and comprehensive restoration plans, funds continue to be raised from federal sources to develop "a world-class, high quality site for Angel Island Immigration Station that places it in its place in U.S. history."[22]

Although not an entirely forgotten episode of history, the integration of Angel Island Immigration Station into the national historical narrative has been slow. Despite a number of exceptional, scholarly articles originating from members of the Chinese American academic community,[23] the history of oppression and resentment against Chinese people in the exclusion era has yet to dent the universality of such myths as those often invoked by the Statue of Liberty: America as land of refuge, opportunity, and freedom for all.

The immigration station's entry into the national historic landscape allows the site's narrators to question such celebratory (and implicitly white) myths of America as a nation of immigrants by articulating a more nuanced version of immigration history where incongruous perspectives are put forward. Jack Tchen, associate professor of history at New York University, noted,

> This process of remembering can be difficult or even painful, since it may lead us to question our fundamental beliefs and "myths" about ourselves. The process is nevertheless an extremely important and valuable one in that it helps us to learn and grow as individuals and as a society. The potential lessons and learnings from the Angel Island experience are vital to our understanding of the broader American experience.[24]

The process of remembering a "broader American experience," however, is not a straightforward task. Contradictions arise when attempting to incorporate a progressive story into the conventional heritage landscape because success in terms of funding and public approval rest so heavily on packaging the landscape to sit more comfortably with uplifting themes.

In this respect the immigration station landscape can serve to both question and sustain the standard foundational immigration narrative. The task for those producing the immigration station today becomes one centered around destabilizing rather than supporting that narrative's universality. As the state park tours proceed through the buildings and on to the poems, and visitors are asked to imagine themselves as immigrants hearing the chimes of the bell ringing through the fog of a typical day in 1920s San Francisco, the context provided must be one that avoids a generalized celebratory account of an immigrant's success over hardship and instead confronts the reasons for that hardship head-on: the racializing of Chinese immigrants and the consolidation of legitimate American identity as white. The immigration station is a landscape with a potential to dramatically effect the construction and reproduction of identity categories. As well as being materially implicated in how categories of race, class, and citizen were constituted during its operation as a facility to police and enact exclusion, the site today, as heritage, continues to be complicit in shaping those very same categories.

Conclusion

I have focused on one particular racialized landscape, and I must at the same time acknowledge that other landscapes are no less implicated in this dynamic. Certainly, if race, as Don Mitchell contended, is a geographical project[25]—a social category constructed to consolidate claims to space by alienating others from it—then race is implicated everywhere.[26] Thus, although the Angel Island Immigration Station is a bounded and coherent landscape, it is simultaneously a distillation of the many geographies of race we inhabit, practice, and reproduce.

The Angel Island Immigration Station, as a National Historic Landmark, centered on the poetic voices of Chinese immigrants, brings into focus questions about national history and national identity and those who are allowed to claim affiliation with it. Here, the restoration of the immigration station is as much a reclaiming of history as it is a reclaiming of territory. Interpreters at the site employ the landscape in realigning the past not only to make room for Chinese Americans in the national landscape but also to assert how that landscape has so heavily implicated racism and practices of exclusion. Acknowledging this is both necessary and potentially empowering.

Notes

1. Information on the immigration station comes from Gareth Hoskins, "Memory and Mobility: Representing Chinese Exclusion at Angel Island Immigration Station" (Ph.D. diss., University of Wales Aberystwyth, 2005). Historical detail is drawn from archival research using documents from the National Park Service and the California Department of Parks and Recreation as well as ethnographical research conducted between 2001 and 2003.
2. California Department of Parks and Recreation, "Immigration Station Tour Script" (2001), 1.
3. Smiley Jan and Tet Yee, "Collection of Autumn Grass: Volume Collecting Voices from the Hearts of the Weak" (unpublished manuscript, 1932).
4. Ling-chi Wang, "The Yee Version of Poems from the Chinese Immigration Station," *Asian American Review* 3 (1976): 117–26; Judy Yung, "A Bowl Full of Tears: Chinese Women Immigrants on Angel Island," *Frontiers* 2 (1977): 52–55; and Connie Young Yu, "Rediscovered Voices: Chinese Immigrants and Angel Island," *Amerasia* 4 (1977): 123–39.
5. This is a translation of a wall carving found in the detention barrack. The authors remain unknown. Partly because of the ambiguity of the poetry style and the difficulty in capturing its subtlety using English, this poem has many versions. This is the most recent rendition and is taken from Mark Davidson and Lauren Meier, *Olmsted Cultural Landscape Report for Angel Island Immigration Station* (Brookline, MA: Olmstead Centre for Landscape Preservation, 2002), 7.
6. Him Mark Lai et al., *Island: Poetry and History of Chinese Immigrants on Angel Island, 1910-1940* (London: University of Washington Press, 1980), 60.
7. Ibid., 159.
8. Ibid., 161.
9. Frank Hays (1916), cited in "Olmsted Cultural Landscape Report" (early draft, California Department of Parks and Recreation, 2002).
10. Visitation figures have been compiled by the Angel Island Association and refer to 2002.
11. Luigi Lucaccini, "The Public Health Service on Angel Island," *Public Health Reports* (January–February 1996); and Nayan Shah, *Contagious Divides:*

Epidemics and Race in San Francisco's Chinatown (Berkeley: University of California Press, 2001).

12. U.S. Congress House Select Committee, Investigation of Chinese Immigration 298. Quote taken from Erika Lee, *At America's Gates: Chinese Immigration during the Exclusion Era* (Chapel Hill: University of North Carolina Press, 2003), 77.
13. Mr. Lowe, who provided this account for the book, was aged sixteen years in 1939. Him Mark Lai et al., *Island: Poetry and History,* 75.
14. U.S. Government, Bureau of Immigration, *Records of the Immigration and Naturalization Service,* Record Group 287, Annual Reports of the Commissioner General of Immigration, 1910–40. Quoted in John Soennichsen, *Miwoks to Missiles: A History of Angel Island* (Tiburon: Angel Island Association, 2001), 134.
15. Jan Rosen et al., *A Teacher's Guide to the Angel Island Immigration Station* (Tiburon, CA: Angel Island Association in cooperation with Angel Island Immigration Station Historical Advisory Committee, 1983).
16. Him Mark Lai et al., *Island: Poetry and History,* 9.
17. Katherine Bishop, "Interview with Paul Chow," *New York Times,* November 1990.
18. Alethea Yip, "Angel Island Campaign Gets Lift," *Asian Week,* December 1996.
19. Daniel Quan, interview with author, October 31, 2001.
20. Ibid.
21. Felicia Lowe, interview with author, November 11, 2001.
22. Michael Wong and Margaret Kadoyama, *Angel Island Immigration Station Foundation Strategic Plan 2001–2006* (San Francisco: Angel Island Immigration Station Foundation, 2001), 2.
23. Roger Daniels, "No Lamps Were Lit for Them: Angel Island and the Historiography of Asian American Immigration," *Journal of American Ethnic History* 17 (1997): 3–18. See also Him Mark Lai, "Island of Immortals: Chinese Immigrants and the Angel Island Immigration Station," *California History* 57 (1998): 88–103; and Connie Young Yu, "Rediscovered Voices."
24. *Angel Island Immigration Station Visioning Workshops Report* (Berkeley, CA: MIG, 1999), 8.
25. Don Mitchell, *Cultural Geography: An Introduction* (Oxford: Blackwell, 2000), 230.
26. See also David Delaney, "The Space That Race Makes," *Professional Geographer* 54 (2002): 6–14.

CHAPTER 6

The Picture Postcard Mexican Housescape: Visual Culture and Domestic Identity

DANIEL D. ARREOLA

The Mexican housescape is a house and its immediate landscape, a specific form of material cultural communication common to Mexican ethnic neighborhoods or *barrios* in the southwestern United States. The housescape is more than a house yet less than a landscape of the size usually studied by geographers and others.[1]

Mexican housescapes have been documented and studied chiefly in contemporary cities of the borderland Southwest.[2] The housescape represents a domestic cultural landscape, rooted in Spanish architectural heritage yet melded to Native American building traditions of Mexico and the greater Southwest.[3] Today, the housescape remains a symbolic landscape in Mexican American communities, created and recreated, often unconsciously, as a representation of a cultural ideal.

This chapter analyzes Mexican housescapes illustrated in popular historic postcards. During the early twentieth century, postcards presented a stereotyped view of Mexican Americans as residing in primitive dwellings and typically in rural habitats. That stereotyped representation contributed to a common primitive image of Mexican ancestry people in the United States. That popular imagery became a factor in American society's view of Mexican American people, place, and domesticity.

More specifically this chapter suggests the process by which postcard representations of the historic Mexican housescape became institutionalized as a media image and sheds light on several questions relevant to Mexicans as an ethnic group in American society. Postcards were part of popular media during the first half of the twentieth century, and images attached to these items of personal communication became widely disseminated. By their very popularity postcards shaped attitudes about people and place. The picture postcard promised to capture and depict everyday life, even though the images presented might bear little resemblance to the lives represented. This way of seeing influenced how society perceived Mexicans and contributed to a pejorative view of this ethnic group in America.

In this chapter I describe first how picture postcards captured this traditional landscape visually according to types of dwellings and by regional identity. Then I compare the visual image of the Mexican housescape presented in postcards to the ethnic literature about Mexican domestic environments to corroborate that a particular image was constructed about the group in the twentieth century. Finally, I argue that the postcard image of the Mexican housescape reinforced the view of Mexicans as primitive and rural, although most Mexicans in America during the period in which the postcards were current resided in central cities and on the margins of urban places rather than in rural districts. In this manner, we see how an ethnic group is both socially marginalized and spatially disenfranchised.

Picture Postcards

The history of postcards is more recent than that of photography, and picture postcards, or view cards as they have been called, first appeared in Europe during the last quarter of the nineteenth century. These early postcards depicted local views and scenes on the front, or message side, of the card; the backside of the card was, restricted by law, for the mailing address only. After 1902 in Europe and 1907 in the United States, divided-back postcards came into being. Divided backs allowed for message and mailing address on the back of cards while the front was entirely devoted to a view.[4]

During the early postcard era in the United States, two companies came to dominate commercial production: Detroit Publishing and Curt Teich. Detroit Publishing is known to have printed postcards from seventeen thousand different images between 1895 and 1935.[5] By the 1910s Curt Teich sold some 150 million postcards annually, mostly view cards of scenes in the United States.[6]

Although corporate postcard publishers dominated national production, independent postcard producers operated in towns and cities across the country. The Eastman Kodak Company, for example, issued postcard-size photographic paper that could be used to print directly from a negative. Soon amateur photographers as well as independent professionals began to produce postcards.[7]

Robert Runyon (1909–68), a Brownsville, Texas, photographer, is an example of an independent postcard entrepreneur. As a local professional photographer, Runyon produced his own postcards and also contracted to companies to convert his photos to postcards. He then made arrangements with local drug and cigar stores and other small retailers to sell his cards. In addition, regional distributors with curio stores in San Antonio and Houston bought several thousand of Runyon's postcards, which included domestic scenes and daily life views of Mexicans in Brownsville and the lower Rio Grande Valley.[8]

Walter Horne (1883–1921), an El Paso, Texas, photographer, is another example of an independent postcard producer. Horne's popular views included scenes of Mexican American life in El Paso and complemented his popular photo postcards of the Mexican Revolution in nearby Ciudad Juárez. In a 1914 correspondence, Horne indicated that he produced thirty thousand postcards that year, with shipments to New York City and Los Angeles.[9]

Postcards produced popular imagery about places before the widespread availability of personal cameras. Because postcards by convention exhibited the local, they can be an excellent source of historic views about a place. Ironically, postcard photographers sought the unique in the landscapes they documented, but typically they tended to capture the ordinary.[10] Representation of the ordinary or vernacular landscape gives postcard imagery great utility in historical geographic research.

Unlike conventional photographic imagery about place that one might excavate from an archive or from official sources, postcards by their nature as mass commercial products were highly accessible.[11] Accessibility meant that this form of imagery became stamped into popular consciousness. In this manner, the identity of Mexicans in the United States was shaped and influenced by the popular postcard.

Mexican Housescapes in Postcards

Writings about the built environment traditions of the Southwest's Mexican ancestry people are surprisingly scarce. This is so despite the general acknowledgment by scholars that Spanish and Mexican communities in the region were pioneering settlements of an expansive Hispanic cultural influence that flowed north from Mexico.[12] Consensus is that early colonizers and their settlements largely followed an age-old building tradition, imported from Iberia via Mexico and modified according to local conditions. Dwellings were constructed of earthen materials such as mud bricks, or wattle, daub, and thatch that resulted in simple vernacular structures popularly known as adobes and *jacales* (shacks), respectively.

Structures in many communities in the Mexican borderland of the United States were initially constructed of these traditional materials, including those in Los Angeles, Tucson, Laredo, and San Antonio. Once Anglo-American populations settled in these communities, building styles were transformed by the use of exotic imported materials and other cultural and regional construction traditions. Nevertheless, in the quarters of these cities that persisted as Mexican districts, traditional building practices survived, especially in poorer communities and in immigrant neighborhoods. As Mexican ancestry residents in these cities relocated from traditional quarters to mixed social areas and suburban districts, Mexican Americans came to occupy dwellings that were similar to dwellings occupied by residents of like social classes.

To evaluate how Mexicans in American communities were represented in postcard imagery, I make use of my collection of historic postcards. My collection includes thousands of historic postcards of various topics, chiefly of towns and places in the borderland Southwest and northern Mexico. One category of my archive contains 213 historic postcards that illustrate Mexican American residential spaces or housescapes. The postcards in this set date from the early twentieth century to the 1940s, but most are from the first two decades of the twentieth century.

I classified the 213 postcards by the dominant type of Mexican housescape construction that is represented in each image. Dwelling types and their frequency include the following: *jacal* (109), adobe (97), stone (6), and stucco (1). More than 96 percent of the dwellings illustrated in these postcards are *jacales* and adobe structures; some 51 percent alone are *jacales* and 46 percent are adobes (see, for example, Figures 6.1 and 6.2, respectively). Dwellings of stone and stucco are much less common than *jacal* and adobe construction in the postcard views of Mexican housescapes.

Regional identification of postcards can be achieved by inspecting the printed place locations on the image, the postmark, and the message written on the card, although each of these can be unreliable without careful inspection of the view. For example, postcard printers were not beyond pirating images and reissuing the postcard with a different place location from the original image. A postcard without printed identification could also be purchased in one location and mailed from another, thus making accurate identification difficult. Experience in collecting and comparing

Figure 6.1 Mexican *jacal,* Mercedes, Texas. Albertype print postcard, c. 1907–15. Arreola Postcard Archive.

postcard images and knowledge of the regional and cultural historical context gives some advantage in identification, but other times regional identity is impossible to determine.

Some 110 of the 213 postcards used in this study—52 percent—can be identified as views of Mexican housescapes in Texas. Roughly 49 postcards, or 23 percent of the total, can be confidently identified as locations in New Mexico. Some 26 postcards, only 12 percent, and 11 postcards, just

Figure 6.2 Mexican adobe dwelling, El Paso, Texas. Photographic postcard, c. 1910–20. Arreola Postcard Archive.

5 percent, are locations in California and Arizona, respectively, and only 6 postcards are locations in Colorado. Eleven postcards could not be identified by location.

This regional representation of Mexican housescapes largely accords with the pre-World War II distribution of the Mexican population in the United States. Until 1950 Texas counted the greatest concentration of Mexican ancestry people in the country, both foreign born and native born.[13] California's present dominance as the state with the greatest Mexican ancestry population extends only slightly more than two generations, and Arizona and New Mexico each have counted fewer Mexicans over the past century. The disproportionate representation of New Mexico in this postcard imagery relative to its historically smaller population might be accounted for by the strong association of that state with Hispanic heritage.[14] In 2000 New Mexico was the most Hispanic/Latino state in the nation—42 percent of its population declared this ancestry—and that condition has persisted since New Mexico became a state in 1912.

Significantly, these postcard images typically included families or family members, often barefoot children, barrels, steel pots, tables with household cookery, laundry, domestic livestock, wagons, ramadas, and corrals in housescape views. The repetition in postcards of this domestic scene helped cement the image of Mexican Americans as living in primitive surroundings.

Social Context of Mexican Housescapes

Representations of Mexican housescapes in picture postcards reinforced a stereotype about Mexican Americans and domestic living. The overwhelming message projected in these postcard views is one of primitiveness. The dwellings appear ramshackle, the individuals posed in the images are often poorly dressed, and the material possessions are suggestive of poverty (see Figure 6.3). This is not to imply that poverty was unknown or unusual among Mexicans in the United States during the early twentieth century. The point here is to understand how the image of poverty reinforced by the postcard view equated with society's stereotype of this condition as characterizing all Mexican residential living.

During the 1920s and 1930s, a popular discussion in the southwestern press and literate circles was the so-called Mexican problem. According to Carey McWilliams, who examined the literature of that era, the problem "apparently consists in the sum total of voluminous statistics on Mexican delinquency, poor housing, low wages, illiteracy, and rates of disease."[15] The problem was chiefly seen as a consequence of Mexican immigration during the early part of the century and the reactions of local communities to the need to create social services for Mexicans during the Great

Figure 6.3 Mexican hut, Tucson, Arizona. Housescapes presented in postcards typically included families in front of modest dwellings along with material trappings, suggesting poverty. Albertype print postcard, c. 1905. Arreola Postcard Archive.

Depression. The Mexican problem as something internal to Mexican culture was never accepted without dissent by certain academicians of the era, and even to the present, social scientists continue to argue the historical implications of these early assertions for Mexican American culture.[16] It might even be asserted that this problem is with us still in the guise of contemporary debates about Latin American immigration, which is chiefly emigration from Mexico.

Especially dramatic evidence of this historic problem was the visible presence of poor housing in cities with Mexican populations. In Texas, Anglos frequently equated Tejanos (Texas Mexicans) with primitive dwellings.

> Texans still considered *jacales* symbolic of Tejano backwardness, irresponsibility, and noninventiveness. Unlike other intimate aspects of Tejano society which were shielded from public view, the *jacal* stood out conspicuously, exposed as a special item or artifact for study. Instead of seeing it as a type of housing made from available materials and one which answered the needs of poor people in an effective way, Anglos disparaged it. The grass and straw roofs, the mesquite walls, the clay or mud floors, the makeshift furniture, and other aspects of the domicile came in for ridicule. Anglos pointed to it as an object of primitivism and backwardness, not as a product of the Mexican capacity for improvisation.[17]

These notions of primitiveness became entangled with ideas about hygiene, and Mexicans came to be associated with dirtiness (see Figure 6.4).

Figure 6.4 Mexican types, boys of the village, Aransas Pass, Texas. Simple clapboard housing made of salvaged scrap lumber was the modern version of the historic Mexican *jacal.* Photographic postcard, c. 1910. Arreola Postcard Archive.

A federal survey conducted in Texas during the 1930s found that the typical houses of Mexicans

> were unpainted one- or two-room frame shacks with single walls, dirt floors, one or two glass windows, and outdoor toilets. The poorer houses were patched together from scraps of lumber, old signboards, tarpaper, and flattened oil cans. Some of the families had no stoves, and the women cooked outside over open fires or, when the weather made it necessary, inside in open washtubs.[18]

In California, social surveys reported about the Mexican shacklike homes built of scrap lumber, old boxes, and other salvage that were typically located on the outskirts of towns.[19] Misunderstood in the public eye, however, was that the primitiveness that was largely associated with a rural lifestyle was, in fact, an urban living pattern. Most Mexicans during this era were urban residents, albeit on the fringes of towns, and other times they were in *barrios* in a central city.[20]

In describing this domestic scene during the 1940s, McWilliams called the phenomenon a "*colonia* complex."

> Scattered throughout Southern California outside Los Angeles are, perhaps, 150,000 to 200,000 Mexicans and Mexican Americans, for the most part immigrants or the sons and daughters of immigrants. … Most of these people—perhaps eighty percent of them—live in "colonies" or *colonias* which vary in size from a

> cluster of small homes or shacks to communities of four, five, six, eight and ten thousand people.[21]

McWilliams described the character of residences in one *colonia* on unincorporated county land on the outskirts of Upland, a suburban community east of downtown Los Angeles and close to San Bernardino.

> With as many as three shacks to a lot, the structures are unpainted, weatherbeaten, and dilapidated. The average house consists of two or three rooms and was built of scrap lumber, boxes and discarded odds-and-ends of material. Ten, twenty, and thirty years old, the houses are extremely clean and neat on the inside ... all the homes lack inside toilets and baths and a large number are without electricity.[22]

The combination of a popular literature about the Mexican problem and the persistent production of postcard imagery that stereotyped the exterior living conditions in Mexican communities created an overpowering sense of Mexicans as primitive and poor. Picture postcards continued to represent Mexican domestic life through images of *jacales* and adobes into the 1930s and 1940s.

In central Los Angeles, only a few miles from the civic center and downtown, were the Mexican *barrios* of Chávez Ravine (where the Los Angeles Dodgers baseball stadium was built in 1962). When Don Normark first photographed these communities in 1948, the place was a landscape of chiefly simple wood-frame homes with tidy yards, strung along the hillsides and flats of the area.

> One rare clear day in November 1948 I was looking for a high point to get a postcard view of Los Angeles. I didn't find that view, but when I looked over the other side of the hill I was standing on, I saw a village I never knew was there. Hiking down into it, I began to think I had found a poor man's Shangri-la. It was mostly Mexican and certainly poor, but I sensed a unity to the place, and it was peacefully remote ... the people of Chávez Ravine lived lives that were a bit more open than those in more conventional American neighborhoods. More of life happened outside their homes, in public, where the stranger's camera could see.[23]

In many respects, the dwellings found in the *barrios* here were no different from those found in many areas termed "Little Mexico" or "Old Mexico" across the Southwest, places inhabited by poor Mexicans (see Figure 6.5). The exterior represented in views of these Mexican housescapes

Figure 6.5 Old Mexico, Pueblo, Colorado. Mexican *colonia* on the edge of a steel mill town. Color print postcard, c. 1910–15. Arreola Postcard Archive.

were often unkempt and ramshackle, but the interiors, as McWilliams found, might be exceptionally tidy.

Yet, overwhelmingly, the postcard image of dwellings in the Mexican American domestic scene was of *jacales* and adobes, structures that were seen by non-Mexicans as primitive. In these views, as Normark noted, life was indeed lived outside as much as it appeared to be lived indoors. To many eyes this exterior housescape might be judged untidy. Extremely rare was any attempt to represent by postcard image Mexican houses that appeared comparable to the formal housing of other urban or suburban Americans, residential spaces where increasing numbers of Mexicans lived.

Visual Culture and Constructed Domesticity

The picture postcard image of the Mexican housescape has been a critical factor in constructing a visual culture about this ethnic group. Visual culture assumes that images do something; they are not neutral. Rose, among others, argued that visual images can be powerful and seductive in their own right, and they have a social impact. That impact is compounded if the image exists in conjunction with other kinds of representation such as text or narrative.[24]

Visual culture insists that images and their look are only part of how social construction emerges; how images are looked at can be equally significant to their impact.[25] The picture postcard is both a form of visual culture representing something and a particular way of seeing.[26] The Mexican housescape represented in picture postcards created a particular domestic view of this subculture, and the way that view came to be seen or consumed shaped society's view of Mexican Americans.

Ethnic group perception as a concept is socially constructed through historical experience, and, therefore, every ethnic stereotype has a historical geography.[27] The popular image of the Mexican American is a product of ethnic narrative construction, chiefly in the nineteenth and early twentieth centuries, but the constructed visual culture of this group is attributable, in large measure, to the picture postcard era in the early twentieth century. The postcard representation was an extension of the stereotyped images of Mexicans in literature and later in cinema.[28]

If we return to Don Normark's observation about Mexican American lives lived outside, we engage a telling clue to the Anglo perception of the *barrio* landscape and ultimately the Mexican housescape as witnessed in American society. During the first half of the twentieth century, many residents of *barrios* across the Southwest emigrated from rural and village Mexico.[29] In those habitats, living outside meant a certain amount of material would accumulate around dwelling spaces. Anglos in the Southwest who were domesticated to a formal eastern or midwestern landscape aesthetic might see these accumulations as clutter.[30] Not appreciated was that a cluttered Mexican housescape in urban and suburban America was more a consequence of residents who were in transition between rural and urban settings than the interpretation of Mexicans as primitive or dirty. Thus the perception that outsiders might have about the Mexican housescape as a result of poverty might also be seen as a legacy of a homeland rural experience where space was

Figure 6.6 Mexican residences, Ajo, Arizona. A rare postcard view of a conventional Mexican dwelling in a company copper mining town. Photographic postcard, c. 1910–15. Arreola Postcard Archive.

abundant and materials could pile up. In the tighter spaces of the city or on the margins of cities, what was once accumulation spread over a rural ranch compound becomes clutter in the *barrio* landscape and thereby is perceived as primitive, and thus captured in the visual image of the postcard Mexican housescape.

Nevertheless, the primitiveness of the Mexican housescape as depicted in postcards fit and reinforced the pejorative image created of Mexicans in the United States in the early twentieth century. Postcard photographers specifically selected these scenes, not images of more conventional urban housescapes that would have been in keeping with the Mexican American housing of the era (see Figure 6.6). The postcard representation thus became the stereotyped view of all Mexicans living in the United States, and that selected visual culture helped construct a dominant if distorted view of its domestic environment and, by extension, of Mexicanness.

Notes

1. James S. Duncan and David Lambert, "Landscapes of Home," in *A Companion Cultural Geography*, ed. J.S. Duncan, N.C. Johnson, and R.H. Schein (Malden, MA: Blackwell, 2004), 382–403.
2. William F. Manger, "The 'Idealized' Mexican American Housescape," *Material Culture* 32 (2000): 1–36; and Daniel D. Arreola, "Mexican American Housescapes," *Geographical Review* 78 (1988): 299–315.
3. Nina Veregge, "Transformations of Spanish Urban Landscapes in the American Southwest, 1821–1900," *Journal of the Southwest* 35 (1993): 371–460; Mario L. Sánchez, *A Shared Experience: The History, Architecture and Historic Designations of the Lower Rio Grande Heritage Corridor* (Austin: Texas Historical Commission, 1991); Robert C. West, "The Flat-Roofed Folk Dwelling in Rural Mexico," in *Man and Cultural Heritage: Papers in Honor of Fred B. Kniffen*, ed. H.J. Walker and W.G. Haag, *Geoscience and Man 5* (Baton Rouge: Louisiana State University, 1974), 111–32.
4. Frank Staff, *The Picture Postcard and Its Origins* (New York: Praeger, 1966).
5. Nancy Stickels Stechschulte, *The Detroit Publishing Company Postcards* (Big Rapids, MI: Nancy Stickels Stechschulte, 1994).
6. George Miller and Dorothy Miller, *Picture Postcards in the United States, 1893–1918* (New York: Clarkson N. Potter, 1976).
7. Hal Morgan and Andreas Brown, *Prairie Fires and Paper Moons: The American Photographic Postcard, 1900–1920* (Boston: David R. Godine, 1981).
8. Frank N. Samponaro and Paul J. Vanderwood, *War Scare on the Rio Grande: Robert Runyon's Photographs of the Border Conflict, 1913–1916* (Austin: Texas State Historical Association, 1992).
9. Mary A. Sarber, "W.H. Horne and the Mexican War Photo Postcard Company," *Password* 31 (1986): 5–15, 46.

10. John A. Jakle, *The American Small Town: Twentieth-Century Place Images* (Hamden, CT: Archon Books, 1982).
11. Daniel D. Arreola, "The Fence and Gates of Ambos Nogales: A Postcard Landscape Exploration," in *On the Border: Society and Culture between the United States and Mexico,* ed. A.G. Wood (Lanham, MD: SR Books, 2004), 43–79; and Joan M. Schwartz and James R. Ryan, eds., *Picturing Place: Photography and the Geographical Imagination* (London: I.B. Tauris, 2003).
12. David J. Weber, *The Spanish Frontier in North America* (New Haven, CT: Yale University Press, 1992); and Gilbert R. Cruz, *Let There Be Towns: Spanish Municipal Origins in the American Southwest, 1610–1810* (College Station: Texas A&M University Press, 1988).
13. Thomas D. Boswell, "The Growth and Proportional Redistribution of the Mexican Stock Population in the United States: 1910–1970," *Mississippi Geographer* 6 (1979): 57–76.
14. Daniel D. Arreola, ed., *Hispanic Spaces, Latino Places: Community and Cultural Diversity in Contemporary America* (Austin: University of Texas Press, 2004); and Chris Wilson, *The Myth of Santa Fe: Creating a Modern Regional Tradition* (Albuquerque: University of New Mexico Press, 1997).
15. Carey McWilliams, *North from Mexico: The Spanish-Speaking People of the United States* (New York: Greenwood, 1968), 206.
16. Arnoldo De León and Kenneth L. Stewart, *Tejanos and the Numbers Game: A Socio-historical Interpretation from the Federal Censuses, 1850–1900* (Albuquerque: University of New Mexico Press, 1989); and John R. Chávez, *The Lost Land: The Chicano Image of the Southwest* (Albuquerque: University of New Mexico Press, 1984).
17. Arnoldo De León, *They Called Them Greasers: Anglo Attitudes toward Mexicans in Texas, 1821–1900* (Austin: University of Texas Press, 1983), 30.
18. David Montejano, *Anglos and Mexicans in the Making of Texas, 1836–1986* (Austin: University of Texas Press, 1987), 227.
19. William Deverell, *Whitewashed Adobe: The Rise of Los Angeles and the Remaking of Its Mexican Past* (Berkeley and Los Angeles: University of California Press, 2004); and Richard Griswold del Castillo, *La Familia: Chicano Families in the Urban Southwest 1848 to the Present* (Notre Dame, IN: Notre Dame University Press, 1984).
20. Daniel D. Arreola, *Tejano South Texas: A Mexican American Cultural Province* (Austin: University of Texas Press, 2002); Albert Camarillo, *Chicanos in a Changing Society: From Mexican Pueblos to American Barrios in Santa Barbara and Southern California, 1848–1930* (Cambridge, MA: Harvard University Press, 1979); Arnoldo De León, *Ethnicity in the Sunbelt: A History of Mexican Americans in Houston* (Houston, TX: University of Houston, 1989); Mario T. García, *Desert Immigrants: The Mexicans of El Paso, 1880–1920* (New Haven, CT: Yale University Press, 1981); Robert Lee Maril, *Poorest of Americans: The Mexican-Americans of the Lower Rio Grande Valley of Texas* (Notre Dame, IN: University of Notre Dame Press, 1989); Arthur J. Rubel, *Across the Tracks: Mexican-Americans in a Texas City* (Austin: University of Texas Press, 1966); George J. Sánchez, *Becoming Mexican*

American: Ethnicity, Culture and Identity in Chicano Los Angeles, 1900–1945 (New York: Oxford University Press, 1993); and Thomas A. Sheridan, *Los Tucsonenses: The Mexican Community in Tucson, 1854–1941* (Tucson: University of Arizona Press, 1986).

21. Carey McWilliams, *North from Mexico*, 217.
22. Ibid., 218.
23. Don Normark, *Chávez Ravine, 1949: A Los Angeles Story* (San Francisco: Chronicle Books, 1999), 11.
24. Gillian Rose, *Visual Methodologies: An Introduction to the Interpretation of Visual Materials* (London: Sage Ltd., 2001); and Deryck W. Holdsworth, "Landscape and Archives as Texts," in *Understanding Ordinary Landscapes*, ed. P. Groth and T.W. Bressi (New Haven, CT: Yale University Press, 1997), 44–55.
25. R.H. Schein, "Representing Urban America: 19th-Century Views of Landscape, Space, and Power," *Environment and Planning D: Society and Space* 11 (1993): 7–21.
26. M. Crang, "Envisioning Urban Histories: Bristol as Palimpsest, Postcards, and Snapshots," *Environment and Planning A* 28 (1996): 429–52; Naomi Schor, "*Cartes Postales:* Representing Paris 1900," *Critical Inquiry* 18 (1992): 188–244; Paul J. Vanderwood and Frank N. Samponaro, *Border Fury: A Picture Postcard Record of Mexico's Revolution and U.S. War Preparedness, 1910–1917* (Albuquerque: University of New Mexico Press, 1988); and Gordon Waitt and Lesley Head, "Postcards and Frontier Mythologies: Sustaining Views of the Kimberley as Timeless," *Environment and Planning D: Society and Space* 20 (2002): 319–44.
27. Kay Anderson, "Chinatown as an Idea: The Power of Place and Institutional Practice in the Making of a Racial Category," *Annals of the Association of American Geographers* 77 (1987): 580–98.
28. Arthur G. Pettit, *Images of the Mexican American in Fiction and Film* (College Station: Texas A&M University Press, 1980).
29. Lawrence A. Cardoso, *Mexican Emigration to the United States, 1897–1931: Socio-Economic Patterns* (Tucson: University of Arizona Press, 1980).
30. Melvin E. Hecht, "The Decline of the Grass Lawn Tradition in Tucson," *Landscape* 19 (1975): 3–10.

CHAPTER 7

Race, Class, and Privacy in the Ordinary Postwar House, 1945–1960

DIANNE HARRIS

Between 1945 and 1960, thousands of Americans purchased a home of their own for the first time. Buying small houses that averaged one thousand square feet in size in newly constructed suburbs, they exercised their perceived rights to ownership of a single-family home, and in the process they began to reconfigure and affirm their identities. The ordinary house—whether a plywood or stuccoed ranch house, a concrete block two-bedroom unit built on a concrete slab, or a split-level (to name a few types)—served as an important framework for racial, class, and ethnic assimilation for a large number of Americans in the immediate postwar period (see Figure 7.1). Those identified as "white" made up the vast majority of new home buyers because others were largely excluded from suburban housing through the racist practices of government lending programs, real estate steering, and restrictive covenants.[1] New houses were produced primarily for a generically conceived "white" audience of presumed middle-class status. In this way the industries attendant to domestic building and design became complicit in the formation of good Americans—which in the context of the postwar era meant implicitly "white" Americans—out of every new home owner in the nation. These houses supported and aided the refashioning of a new personal and family identity that was, whenever possible, constructed with little or no reference to a past tradition.

Figure 7.1 An ordinary postwar house, Urbana, Illinois. Photo by Dianne Harris.

As recent scholars have noted, and as I expand on here, "home" must be understood not only as a site of safety and shelter "but also as the site of an enormously creative process—the formation of a cultural identity."[2]

This chapter examines new and increasing concerns for personal and family privacy as a determining factor in the design of domestic interior and exterior spaces in the postwar period. I focus exclusively on the desire for privacy in relation to outsiders, although I should acknowledge that privacy *within* the house was equally a concern shared by home owners, designers, builders, and critics. Privacy, like race, is historically contingent and culturally constructed.[3] It is not universally privileged, nor is it monolithically constituted over time and space. But in the United States after 1945, concerns for maintaining privacy in the domestic realm became a pervasive theme in the house design and construction literature. Books and magazine articles aimed at both the middle majority and an audience who could afford an architect-designed house repeatedly emphasized the need to exclude the outsider's gaze and to reduce interior familial frictions through proper design for privacy, while maintaining modernism's various aesthetic codes.

The increased use of expansive glazing in postwar homes certainly brought privacy concerns into sharper focus. Although custom-designed homes often included extensive areas of glazing, ordinary houses usually featured a single, prominent picture window or sliding-glass patio doors, or both. The increased transparency signaled aesthetic modernity and therefore class distinction but also resulted in a lack of internal privacy, especially if the glazing appeared on facades that faced the street. The concern that one might "live like a goldfish in a bowl," with all of the family's activities observable by strangers and neighbors, appears repeatedly as a plaintive refrain in the literature of the period.[4] Despite such anxieties, most postwar Americans lived more privately than ever before, with more seclusion among family members and more separation from neighbors than experienced by previous generations, yet evidence indicates that they worried far more about it.[5] It would be difficult to find a single book or article on general house planning and design from the period that ignored the topic, and most featured privacy as a focus.[6] Certainly, privacy constitutes an unremarkable, even quotidian, planning concern, one that any pragmatist can understand. But the intensity of focus this issue universally received in the literature signals a deeper significance. These published works directly affected house construction nationwide, and we can read them as more than just a register of elite values. The messages about domestic privacy shaped suburban forms and suburban housing for decades to follow because they resonated with a range of concerns held in currency throughout the nation.

Why was privacy so important to postwar American home owners? What did privacy symbolize for them, and how was it achieved? How did the pervasive concern for attaining a private residential world change the design and construction of ordinary postwar houses? What did the published discourse on privacy signal about contemporary American values? In this chapter I demonstrate that owning a single-family detached house with its own private, fenced garden symbolized not just security from outsiders who might threaten home and family but also the security of confirmed membership in the white, middle-class, American majority. The absence of residential privacy was a key feature of the tenements and crowded apartments that signaled prewar, immigrant-based, blue-collar, and lower-class lifestyles—something many Americans wanted to leave behind as they fled to new developments in the suburbs. No matter how small the new, suburban house, it was still a house of one's own, on an individual and distinctly defined lot, separated from the noises, smells, and activities of extended family members, neighbors, and street life that recalled inner-city, prewar dwelling patterns.

Privacy here is viewed as a code for excluding others, whether neighbors or strangers, such that the privacy discourse is exclusionary. Design for privacy is therefore design for exclusion and is about spatial purification. As David Sibley wrote, this exclusionary drive leads to the "never-ending invitations to consume further the privatization of the family, which is closed off from the outside world. Life beyond the home enters the private sphere through stereotyped images, conveyed by videos, television commercials and similar media messages."[7] Privacy is associated with notions of a pure self, a pure identity, and a pure family, and all of these must remain unsoiled by outside influences. Privacy therefore necessitates maintaining rigid boundaries, which ultimately become exclusionary. Fears about the loss of privacy must also be seen as a "fear of pollution" that comes from the others' actions.[8] Moreover, because the right to private property ownership and control in the United States has been historically reserved largely for "whites," the simple fact of home ownership symbolizes the exclusion of racial minorities. As Kevin Fox Gotham noted, privacy starts at the property boundary and is related to the deed of ownership and an entire ideology that is encapsulated within it, such that the discussions of private property rights must be viewed as a language of exclusion.[9]

In the discourse on privacy, race and class are as inextricably intertwined as they always are, each the operative modality of the other.[10] The rhetoric of taste, so frequently deployed by architects, designers, builders, critics, and tastemakers, must be understood as equally the rhetoric of both racial and class formation. The published texts on privacy cited throughout this chapter were frequently linked to issues of good and bad taste and to an aesthetic frame, and as such, the literature on taste preferences must also be understood as the deployment of exclusionary practices.[11]

The remaining portions of this chapter interrogate issues of race and class as they were imbricated in discussions of and concerns about privacy in sociological, social critique, and design literatures. The suburban house and garden became the explicit focus of sublimated race and class politics in America in the postwar era. Design professionals, tastemakers, and cultural critics called for specific interventions in postwar housing construction, practices, and styles, using books, magazines (popular, shelter, and trade publications), television, and radio to broadcast their ideal. The houses located throughout the suburban landscape of the United States serve as the built, material record of the popular acceptance of the persuasion to privacy, their carefully placed walls, fences, and hedges a testament to the widespread acceptance of the exclusionary ideal.

Privacy, Individuality, and American Identity

In the postwar decades, Americans became more fully aware that their intimate lives could be exposed without their knowledge through the use of invisible, albeit mechanical, eyes and ears, and this certainly contributed to the growing concerns for privacy. New technologies created and implemented during the war were now being put to domestic purposes, resulting in a rash of anxieties about surveillance mechanisms. Surveillance technology was prevalent in law enforcement and in private investigation in the 1940s, but it didn't really enter the public consciousness until after 1955, when a growing public anxiety related to the endangerment of personal and public privacy emerged.[12] To understand such anxieties as they translated to the domestic realm, one need only recall John Cheever's short story "The Enormous Radio," the tale of a mechanically malfunctioning receiver that inexplicably reveals the intimate lives of an entire New York apartment house to its voyeuristic owner.[13] The invention of parabolic microphones, small wireless resonator radio transmitters, and cameras—which were sometimes referred to as "television eyes"—small enough to be hidden in heating ducts or lighting fixtures stimulated broad civic discussion about wiretapping between 1953 and 1955. The 1953 hearings on "Wire-tapping for National Security" carried out by a subcommittee of the House Judiciary Committee were accompanied by coverage in the popular press that featured "the arrival of 'Buck Rogers' technology," so that public reaction was typically contradictory, embracing space-age science while rejecting its implications for maintaining their private lives.[14]

Americans also worried about the use of subliminal suggestion delivered through various media but specifically through television, particularly after about 1957 when all the major news services were carrying stories about the topic. Significantly, Vance Packard's *The Hidden Persuaders* also appeared in that year, a book that warned readers about a range of techniques that could be used to penetrate their psyches, including hypnosis, which the author claimed was being used by corporations to explore the minds of model consumers.[15] Subliminal suggestion involved "the projection of messages by light or sound so quickly and faintly that they are received below the level of consciousness," and the topic became linked to national debates about ethics and the social impact of advertising.[16] An article in the *New Yorker* stated, "[Americans] had reached the sad age when minds and not just houses could be broken and entered" and that "nothing is more difficult in the modern world than to protect the privacy of the human soul."[17] Remembering that both television and popular psychology became widely available during the 1950s, it is no surprise that Americans feared subliminal suggestion. The fears of privacy loss affected every aspect and scale of life, from the outdoors to the indoors and even into the mind of the individual.

Although keeping up with the proverbial Joneses became a postwar national pastime, some sociologists of the period believed that achieving a degree of individuality through private introspection and leisure was essential to the construction and maintenance of class identity and even to the preservation of democracy.[18] A number of sociological studies from the period recorded a growing concern about the effects of mass suburban conformity, and all these publications were linked to the discourse on privacy.

These texts, as Winnie Breines pointed out, were concerned primarily "with the defense of American democracy against the danger of mass movements such as communism and McCarthyism or, some have argued, against the masses themselves."[19] The campaign against communists, both within and without national borders, was indeed central to this anxiety, and Breines noted, "Literal and figurative boundaries were important in the fifties, a period in which distinctions between 'them' (foreigners and deviants) and 'us' flourished."[20] But it could be difficult to recognize some cold war enemies, because communists bore no specific physical attributes. The idea that threatening people, activities, or even germs (especially in the decades of recurring polio epidemics) could be surrounding one's family at any moment and without recognition was deeply troubling. People and ideas that were scary or threatening were therefore to be separated, segregated from everyday life to the greatest extent possible.[21] If communists could be difficult to recognize, racial differences were thought to be easily perceived, and establishing boundaries that excluded those deemed undesirable because of class, ethnicity, or color was viewed as a necessity. The fence achieved this social partitioning at the microscale of the house. At the macroscale of the city, the Federal Housing Authority applied its hand at restrictive zoning, real estate agents engaged in steering practices, and developers exerted influence by enforcing covenants that kept people of color from most suburban neighborhoods.[22]

Cold war technologies, "Red Scare" paranoia, fear of corporate consumer ubiquity, and racial tensions were central to burgeoning concerns for privacy and individuality in the 1950s. Yet many sociologists and social critics focused their fears on suburban life and suburban form rather than on the more complexly formulated realms of politics and culture. Societal overconformity, especially in the suburbs, was their immediate and explicit concern. Their writings called for individuality as achieved through residential privacy as the antidote to conformity. Suburban form, then, became the explicit focus of design critics and professionals, who prescribed residential privacy as central to a distinctive American identity. In these writings and design prescriptions, *individuality* and *privacy* appear as code words for *white* and *middle class* or better. These publications underscored anxieties about class and race that were mediated through the material landscape of postwar suburban housing development. Suburban landscapes bore the weight of these fears and anxieties in the guise of concerns about individuality and privacy.

David Reisman's 1950 novel *The Lonely Crowd* is one such example. Reisman's was a study of national character formation, and he focused on a perceived trend toward postwar overconformity. He shared this concern with Lyman Bryson, who considered cultivating individualism the duty of every American for the preservation of democracy.[23] Reisman encouraged Americans to "break free of their conformist peer-group aspirations," and he sought possibilities for developing an autonomous society. Such studies undoubtedly arose from anxieties related to images from World War II and to the recognized horrors of fascism, which disallowed individuality and diversity. But they were also linked to concerns about the growing homogeneity of suburban life. In the closing section, titled "Autonomy and Utopia," Reisman extolled the virtues of city planners whom he called "the guardians of our liberal and progressive political tradition" and advocated a view of the city "as a setting for leisure and amenity as well as work."[24] The author considered recreation and leisure vital components in the fight against the mass conformity fostered by the workplace and the suburban tract.

Reisman believed that if individuality was to be attained, one had to have privacy as well, because privacy fostered self-expression and inward contemplation, both of which facilitated freethinking. But such freethinking was ultimately linked to democracy, to the American way of life.[25] Backyards and houses, then, became a key to individualization, a means to autonomy and ultimately, it was hoped, to the strengthening of democracy. The key was to increase the amount of leisure time for suburbanites and to help Americans—especially the new middle majority—achieve a degree of distinction, without appearing eccentric or radically different. The balance was crucial: one's house and garden should reflect one's outlook and personality but should conform to a level of embellishment already established in the neighborhood and following the guidelines set out in tastemaking books and journals.[26] As Russel Lynes noted in *The Tastemakers,* "A home of one's own meant a house different from one's neighbors … [a house that had] a semblance of individuality without a trace of eccentricity. … Taste was a quality to be carefully strained, and the court of appeal on all such matters was first a peek into your neighbor's window and then a careful study of the women's magazines."[27]

Lynes underscored the fundamental tension that existed between the desire for maintaining privacy and for cultivating distinction through emulation. Postwar home owners were told to look inward to develop their individual lives, but if they didn't take at least a peek outward into their neighbor's picture window to see what was on display, or into the women's magazines to see what everyone else was admiring, and perhaps purchasing, they couldn't be certain of either their emulative success or their achievement of one-upmanship. To keep the curtains drawn, or to allow the neighbors that moment of voyeurism required by parties on both sides of the glass, became a dilemma of modern domestic life.[28]

Among the well-known publications that made explicit connections between the conformity of suburban residences and the loss of individualism was John Seeley's *Crestwood Heights: A Study of the Culture of Suburban Life;* William Dobriner's *The Suburban Community,* which contains an essay by Philip Ennis on "Leisure in the Suburbs"; William H. Whyte's *The Organization Man;* and John Keats's classic diatribe against suburban living, *The Crack in the Picture Window.* In his essay, Ennis wrote, "Leisure activities, therefore, become an important source of self identification. … Leisure styles are often the basis of self image and subsequently of group membership criteria," thereby forging an explicit connection between leisure and individuality, and its meaning in a larger societal organization.[29] But Keats's arguments against suburban conformity had a much broader impact, and he went the furthest toward painting a bleak picture of a suburbia filled with drone housewives and characterized by homogenous anomie that threatened the democracy. He wrote,

> Mary Drone in Rolling Hills. … Dwelt in a vast, communistic, female barracks. This communism, like any other, was made possible by destruction of the individual. In this case, destruction began with obliteration of the individual house and self-sufficient neighborhood, and from there on, the creation of mass-produced human beings followed as the night the day. … If we are going to live in bedroom neighborhoods, we must either accent our individualities or all go to hell in the same handbasket, and it's as simple as that. In a homogenous community of look-alike houses peopled with act-alike neighbors of identical age groups, there's not too much we can do to improve our lot except accent such small discrepancies as may exist, and lock our differences within our doors to keep them safe. … More insidious and far more dangerous than any other influence, is the housing development's destruction of individuality. … The closer we huddle together, the greater this pressure for conformity becomes. … The physically monotonous development of mass houses is a leveling influence in itself, breeding swarms of neuter drones.[30]

Keats's swarms of drones invokes a subtext that is far more subtle than fears for an imperiled democracy. His "act-alike neighbors" in the housing development were equally upsetting because people and things that looked exactly alike were associated with nonwhite, lower-economic groups. White, upwardly mobile Americans were thought to possess the cultural and symbolic capital that allowed at least a small degree of distinction.[31] Racial stereotypes of the period even negated the individuality of nonwhite

Americans, promoted through the insulting and fallacious notion that all African Americans, or all Asian Americans, looked alike.

This association between house and racial conformity appears in Elizabeth Mock's 1946 publication *If You Want to Build a House*. Mock wrote, "The real basis for house-planning should be the individual, not the group," and she illustrated her assertion with a cartoon depicting Native Americans entering a tepee.[32] For Mock, the tepee was a lower, vernacular form of architecture, one tepee indistinguishable from the next, and therefore a perfect illustration of the lower-economic-class housing her readers hoped to avoid by designing or selecting houses that were both private and distinctive, far from the image portrayed in the Malvena Reynolds song "Ticky Tacky Houses All in a Row." She further asserted that the modern house was particularly significant because, if designed correctly, it held the potential to be "wonderful and vital and deeply individual, inside and out."[33]

Homogenous, cookie-cutter houses were likewise associated with that particular form of lower-economic-class living, the trailer or mobile home, long associated with "white trash" or an off-white form of whiteness.[34] George Catlin and Mary Catlin's 1946 book *Building Your New House* advocated the purchase and use of stock house plans but drew the line at prefabricated trailers or anything that resembled them. They warned their readers that prefabricated or mass-produced houses lacked the individual expression their readers required, referring to the "Wingfoot Home" as "a sort of glorified trailer."[35] The Wingfoot was a prefabricated, compact housing unit that resembled a trailer, but it also offered affordable housing that could be constructed rapidly during a time in which those attributes were desperately needed in the American housing market. Still, the Catlins discouraged their readers from purchasing the Wingfoot or other prefabricated and mass-produced units, because they believed that by purchasing and constructing from stock plans, their readers could obtain "an individual home."[36] A stock plan could, after all, be manipulated to fit its owner's tastes and requirements, therefore providing a degree of distinction that signaled solid membership in the white, middle majority.

The sociological writings, suburban critiques, and design prescriptions are therefore significant for what they reveal about the pervasive ideology of upward class mobility and, though frequently unstated, its links to race. It was far easier and far more acceptable to admonish suburban dwellers to cultivate distinction and privacy for the sake of democracy than for the preservation of an economic ideology that was inherently linked to racial assignment and to the preservation of all-white suburbs.

The design literature also promoted the link between establishing a private domestic realm and creating familial bliss. Kate Ellen Rogers wrote in

The Modern House, U.S.A., "Family atmosphere is conceived as a protective zone in which children can healthily grow to autonomy," and she likewise asserted, "The family with personal values puts the individuality of each member first, stressing personal enjoyment and privacy. This group more than the others valued 'good taste' and were concerned about the design of their homes."[37] Rogers equated families who cared about their privacy, as expressed through design of house and garden, with those who possessed a higher standard of taste and, therefore, were of a higher class than those who did not.

Proper design of house and garden constituted the clear antidote to overconformity, and the shelter magazines, particularly *House Beautiful*, were vociferous in praising the need for houses designed to maximize privacy. Elizabeth Gordon, who served as the editor of *House Beautiful* between 1945 and 1965, was also deeply concerned about societal conformity, and she used the magazine as a forum to advocate freethinking and individuality expressed through design of the home. Gordon chose the magazine's 1952 "Pace-Setter" home to exemplify the so-called American Style she repeatedly advocated.[38] According to Gordon, the house constituted "a relaxed, democratic architecture—a modern house that belongs, yet has an individuality essential to personal culture. Just as it is the essence of Americanism for each of us to develop our differences, so the Pace-Setter, while honoring the general character of the community, arrived at distinction and originality because it freely solved the problems of a unique site and a particular owner."[39] The house struck the requisite and perfect balance for suburban dwellers and served as an ideal example of the elevated class status of the architect-designed home, which few of her readers could actually afford. The American Style modernism that Gordon and her staff repeatedly advocated, then, was a soft or "everyday modernism" that retained comforting signs of the traditional (hipped roofs, familiar materials such as wood and stone) and that was linked to the editor's belief in the importance of autonomy to development of democratic national character.[40] Attaining an American Style home would certainly have held great appeal for a generation of assimilating home buyers that included immigrants who had only recently received citizenship, a "white" racial assignment, and a middle-class identity.

To achieve this identifiably American Style in house and garden design, privacy was absolutely essential. Without privacy, there could be no autonomy, no democracy—and these were closely linked to the idea of individuality and therefore of "whiteness." As Elizabeth Gordon stated in her 1953 speech at the Chicago Furniture Mart,

> The challenge of our time is individualism versus totalitarianism—democracy or dictatorship—and this struggle is on many fronts. Our front, yours and mine, happens to be on the home front. … It is a time of profound spiritual crisis. … The individual is under assault from many sides. … We judge all design for the home in terms of what it offers for the encouragement of individuality, for the development of individual differences, for the provision of privacy and personal creativity, in short, for what it contributes to the humanistic values of a democratic age. … The modern American house—the good modern house … provides privacy for the family from the community, and privacy for individuals of the family from each other. It inspires democratic living by encouraging a personal life.[41]

Because Gordon equated privacy with the development of individuality and democracy, she devoted more pages to articles related to privacy than to any other aspect of modern design.

Gordon's garden editor, Joseph Howland, authored a 1950 piece for *House Beautiful* titled "Good Living Is NOT Public Living," which connected privacy to the American dream of individual home ownership. He wrote,

> We Americans give much lip service to the idea of privacy. We consider it one of the cherished privileges we fought a war to preserve. Freedom to live our own lives, the way we want to live them without being spied on or snooped around, is as American as pancakes and molasses. … The very raison d'être of the separate house is to get away from the living habits and cooking smells and inquisitive eyes of other people … if your neighbors can observe what you are serving on your terrace, your home is not really your castle. If you can't walk out in a negligee, to pick a flower before breakfast without being seen from the street or by the neighbors, you have not fully developed the possibilities of good living.[42]

In this passage, Howland evoked a number of key phrases that would have resonated powerfully with nouveau suburban *arrivistes*. He summons cold war surveillance paranoia in one sentence, then plays on fears of Depression-era conditions related to economic class (memories of living with noises and cooking smells from neighbors) and anxieties about exposure of the body (the negligee seen by neighbors—the "white" body, as a desexualized body, is always concealed) all in one paragraph, making a compelling argument for proper design for privacy.[43] These were precisely the urban conditions *House Beautiful*'s readership of largely suburban home owners had fled, and Howland's argument for the private residential world was cleverly constructed to resonate with his reader's anxieties about their identities.

Designing Privacy

How, exactly, were postwar houses to be designed then? How could privacy from the outside world be properly attained so that the desired identity could be established and affirmed? The primary points of possible visual intrusion were at the property lines and the windows. The front yard was ostensibly public space, and the back was clearly private. Postwar backyards were inherently more private than their predecessors because most development plans of the period did not include service alleys or back lanes running behind the lots, so that the rear garden became less subject to intrusion than ever before.[44] But front yards, though visible to the street, could not retain the public function of urban "stoop culture," or even of earlier suburban "porch culture," because a life lived on the street, in front of the house, signaled both nonwhite and prewar economic conditions and facilitated exactly the kind of public life the sociologists and design critics cautioned against. Front lawns were and still are the most common treatment for suburban front yards, but in the postwar period, social uses for the lawn decreased, so it became perhaps most important as a horizontally disposed green barrier—a means for proclaiming a defensive zone between house and sidewalk or street. Unspoken rules dictated that strangers not walk on another's lawn, and although children sometimes played games on them, they did so more frequently in the protected backyard.[45]

The idea of the backyard as a private family zone secluded from the conforming masses and the eyes of passing strangers held great appeal for numerous reasons. First, the postwar garden accommodated a range of activities that formerly took place away from home. As one author wrote, "Today all the facilities that used to be scattered around the community we now want to exist on our own little piece of land."[46] The garden became a place for "vitamin-conscious moderns" to relax, a new place for housewives to cook on the outdoor grill, a playground for the children, a recreation center for teens complete with stereo system and swimming pool, and an extension of the living room for adult entertaining. As long as the garden was properly furnished with equipment, furniture, sound system, lighting, and climate control devices, the family members need never leave their property to fulfill their recreational needs.[47] The desire to avoid public recreational facilities and spaces in the immediate postwar period was no doubt connected to recurring polio epidemics as well. But the urge toward insularity appears throughout the period in articles that urged readers to turn their ordinary backyard into "Country-Club Living" and to "make home more exciting than anywhere else, canceling the need for seeking family pleasures in private clubs or public beaches."[48] Again, contradiction surfaces: Americans in the 1950s were more mobile than ever before because of the enormous increase in automobile ownership. Yet the design literature insisted that home was better than anyplace else, and the goal was to leave it as seldom as possible. Even the controlled setting of the country club was less desirable than the insularity of one's own suburban backyard.

The currency of the familial myth, coupled with anxieties about outsiders of any kind—whether immigrant aliens, communist invaders, or people of color—caused most authors, designers, and tastemakers to direct their thoughts toward privacy from the outside world. If the family was to be treasured and cultivated, the only dangers that could be acknowledged were those external to the home, because the family was considered and promoted as sacrosanct, its purity to be maintained at all costs.[49] When the Los Angeles-based architect A. Quincy Jones published photographs of his own steel house in the January 1957 issue of *House & Home,* he stated that he implemented an open plan arrangement with floor-to-ceiling curtains used as the only interior dividers, because "*Inside* the house you're always with your family or your friends—*outside* is where you want privacy. That's why we tried to provide as much privacy as we could, with screens, walls, fences and planting."[50] For Jones, as for many members of the architectural and design community, privacy requirements were externally, rather than internally, dictated, and they could be resolved by implementing a range of vertical screening devices.

In the garden, constructing a private world was predicated on good fencing, or at the very least on implementing a dense, well-clipped hedge. Many early suburban developments were designed with continuous backyard spaces that remained uninterrupted by fencing. Some have remained so, though fenced yards became more common as the early suburbs aged. Still, most of the normative residential literature advocated attractive fence or wall design to provide privacy without offending one's neighbors (see Figure 7.2). *House Beautiful*'s January 1949 issue featured an article stating that the secret of a useful backyard was "Privacy from Nosy Neighbors," which was to be attained through careful design of fences, hedges, and plantings that would "discourage over-the-fence talk," block the view of "would-be second-story gazers," and keep out "prying eyes" and "snoopers."[51] By 1960 Elizabeth Gordon found the subject compelling enough to devote an entire issue to "Landscaping and Privacy," asking her readers, "Is Privacy Your Right or a Stolen Pleasure?" Linking politics and domestic design once again, Gordon urged her readers to consider their

“Good fences make good neighbors” Robert Frost

From “Mending Wall”

These fences make privacy, but are not forbidding. Both sides are good-looking, keep their good looks all year without much attention

One of many well-designed, woven fences you can buy ready-made, ready to be attached to your posts. Such a fence reflects less heat than masonry wall, also lets a little breeze through.

Wood overlay and espalier planting enriches a high brick wall —same texture, less unfriendly. Neighbor across the street gets attractive view; you get complete privacy.

Figure 7.2 “Good Fences Make Good Neighbors” is one of many articles on the design and construction of privacy fences that appeared in popular and shelter magazines in the postwar period. Reprinted by permission from *House Beautiful,* January 1950. The Hearst Corporation. All rights reserved.

political commitment to individuality and the right to privacy. She wrote, “Does Your Front Lawn Belong to You—Or the Whole Neighborhood? The United States is split into two factions over this question—an ideological split just as real as the Republican–Democratic divide. Where do you stand? … The issue really boils down to whether or not others have the right to look at or onto your land.” The editor encouraged her readers to stop watching each other and, borrowing a phrase used by the sociologists, asked them to “turn inward.”[52]

Because fencing was key to achieving privacy, she advised her readers to organize their communities to eliminate deed restrictions and covenants that restricted or prohibited fence construction. Although she noted that such restrictions frequently extended to “the kind of people to whom you can sell your house,” she did not elaborate on the problems of racial discrimination in the residential real estate market. But she did advise her readers to take the law into their own hands if conventional organizational efforts failed, writing, "If you can't get around fencing ordinances legally, there are a few ways to avoid them without breaking the letter of the law," and she recommended hedges, trellises, and climbing vines as suitable alternatives to constructing fences. Because postwar suburbs were racially quite homogenous, the fear of the neighbor's prying eyes seems a bit strange. After all, they were almost certain to be "white" eyes. But as the decade of the 1950s progressed, some degree of integration began to occur, and much of it came with high-profile cases that received national publicity.[53] Moreover, the decade saw an unprecedented degree of class and economic mobility that was likewise accompanied by a new fluidity of racial constructs, as Jews and others became increasingly classified as "white folks."[54] Fences, therefore, became increasingly important in a suburban world whose neighbors' identities were less certain than ever before.

Despite its centuries-old association with American home ownership, the "little picket fence" seldom appeared, and its implementation was not advocated, because pickets allowed too much freedom of vision between and over the stakes. Instead, magazines and books provided examples constructed of tall wood planks used in various patterns, concrete block, corrugated plastic panels, and densely woven aluminum screens, among others (see Figure 7.3). The California landscape architect Douglas Baylis published numerous articles on fence designs, such as the one for the April 1950 issue of *Better Homes and Gardens* that featured louvers, wire fences planted with vines, trellis fences, and so on.[55] Some of the designs are strikingly whimsical, such as his privacy screen that incorporated raised plastic sprinklers that both irrigated vegetation and served as a fanciful water feature for children's play. His striped canvas partitions and cantilevered plastic sheeting screens and his checkerboard fences designed of plywood panels alternating with wire mesh were decidedly innovative and modernistic in appearance. Likewise, the experimental "X-100" steel house designed by A. Quincy Jones for the merchant-builder Joseph Eichler in San Mateo, California, had a concrete block wall along the street frontage that served as a garden enclosure, as a boundary containing the children's play area, and as a wall for the father's workshop (see Figure 7.4).[56] The placement of a fortresslike wall to screen the entry and front windows from the street became a commonplace, especially among architect-designed houses of the period. But it also repeatedly appeared as a recommendation for attaining privacy in literature aimed at middle-class home owners, such as *Popular Mechanics* and in the University of Illinois's Small Homes Council circular series. A 1957 article in *House & Garden* titled "Seclusion

Figure 7.3 An example of a corrugated plastic panel used as a rolling privacy screen. Design by Thomas Church. Reprinted by permission from *House Beautiful*, copyright July 1954, p. 79. The Hearst Corporation.

by Design" illustrated the point best, stating, "Behind a Camouflage of Screens and Walls: Blessed Solitude."[57] The house illustrated in the article, designed by the architect Frederick Emmons for his family, is completely closed off from the street by a series of masonry walls, translucent screens, windowless front walls, and a carport street facade (see Figure 7.5).

Furthermore, authors such as Mary Catlin and George Catlin, whose works were directed at a truly middle-class audience of home buyers and builders on restricted budgets, instructed their readers to place most of the large windows and major expanses of glass toward the back of the house to avoid problems with prying eyes.[58] Ordinary postwar houses increasingly did just that. Fewer and smaller windows appeared on the street facade

Figure 7.4 The Eichler "X-100" house, designed by A. Quincy Jones. The concrete block wall provides a privacy barrier for the house along the street facade. Photo courtesy of Elaine Sewell Jones.

over time, and postwar houses increasingly opened up to the back of the lot, especially when a large wall did not conceal the house from the street in front. Placing the kitchen window at the rear of the house, it was reasoned, also allowed mothers to more easily supervise their children's activities in the enclosed backyard without leaving their work in the kitchen, although as many kept the kitchen on the street side of the house to allow a greater integration of living areas and outdoor spaces in the backyard.[59]

In addition to the standard materials—wood, concrete, and concrete block—a variety of new materials, some developed and tested during the war, made inexpensive and attractive screens. By virtue of their very

Figure 7.5 "Seclusion by Design," a house designed by the architect Frederick Emmons for himself, Los Angeles, California. Reprinted by permission from *House and Garden,* January 1957. Condé Nast Corporation. All rights reserved.

newness, they likewise signaled modernity and personal distinction. Translucent corrugated plastic, frosted Plexiglas, Cel-O-Glass (plastic-coated wire mesh), Transite fencing, and translucent Alsynite (which looks like corrugated plastic and was used as an overhead filter) all appeared in the pages of magazines to aid the home owner in his or her selection of screening material. By closing off the outside world from the domestic precinct, distinction could also be achieved simply through visual and physical exclusion. Postwar houses attained an air of familial exclusivity by shutting themselves off from surroundings and neighbors, just as villas and estates of the wealthy had traditionally created an exclusive compound by implementing surrounding walls.

The fortresslike appearance created by privacy walls was also linked to crime rates that rose toward the end of the 1950s and to the increased desire for personal security. But as early as 1947, *House Beautiful* featured a home with a noteworthy attraction: "a peephole concealed in the west wall of the kitchen so that visitors ringing the front doorbell [could] be surveyed before they are admitted."[60] Within ten years' time, security concerns were such that articles featured radio-operated garage doors, like the one that "enables Mrs. Lindsley to stay in her automobile until she is safely within the confines of her home" and the built-in intercom system that permits Mrs. Lindsley to answer the door from the main house."[61] With privacy walls concealing the house from the street, automatic garage doors that ensured closure of the largest opening into the house, and intercoms that monitored visitors, the postwar house became more fortified against the outside world than ever before.

The mania for individual and family privacy, then, changed the look of postwar suburban developments from their precursors, which had been largely unfenced, so that large, unbroken stretches of lawn and garden passed between house lots, giving an impression of a common greenway. Instead, postwar developments became increasingly fenced, broken by the regular rhythm of partitions that separated lot from lot, neighbor from neighbor, family from the street and its occupants. Although some early postwar suburbs, Levittown among them, initially prohibited the use of fences and walls between house lots, they eventually succumbed as well to the aesthetic of privatization that demanded a fenced yard.[62] By 1969, when the renowned landscape architect Thomas Church published *Your Private World,* the formula for creating private, fenced, outdoor gardens was firmly in place.[63] The backyard had become essentially interior space through enclosure, just as the inverse relationship was occurring inside the home: interiors became exteriorized through the extensive use of glass.

Walls that concealed the front of the house from view were important dimensions in attaining exterior privacy, yet they seldom appeared in houses that were not custom designed. They were a hallmark of Eichler homes, which were designed for the middle majority, but, as noted earlier, Eichler hired architects such as Anshen and Allen, A. Quincy Jones, and Fred Emmons to design his tract houses.[64] Stock house plans that could be purchased seldom specified designs for construction beyond the exterior walls of the house, and though readers were urged to employ a landscape architect to design their gardens, many could not afford to do so. The majority of new home owners, therefore, faced the dilemma posed by the ubiquitous picture window—an element that signaled modernity through its extensive glazed surface and allowed the requisite display to neighbors but that was deeply problematic for anyone concerned with privacy.[65] Indeed, the picture window became the target of much criticism from sociologists, home owners, and design critics, who all tended to see it as a problematic element.

According to Sandy Isenstadt, picture windows became popular in the early 1930s, when the glass manufacturer Libby-Owens-Corning began advertising "The Picture Window Idea" in home magazines. Isenstadt's study reveals the conflicts that resulted when a form that had its origins in the ribbon windows of high-style modernism became "demonized as emblematic of pretty much everything wrong in architecture, America, or both." As Isenstadt wrote, "In architectural circles, picture windows became the apotheosis of commercial vulgarization: the subordination of high ideals to crass consumerism."[66] In its favor, the picture window allowed increased amounts of daylight into the home and the promise, if not the reality, of an ever-changing, suburban pastoral view—a view that signified wealth for its links to an Arcadian, romantic past. But as Isenstadt also pointed out, views came to acquire real cash values in the real estate markets of the 1940s, when " 'view' began to appear as a line item on appraisal forms."[67] To be able to claim a view outside one's picture window, then, also signaled wealth in the real terms of market value and therefore became another affirming image of class status. Moreover, small windows signaled the past, and perhaps even impecuniosity, because large areas of glazing had long appeared as a symbol of wealth.

But working against the picture window were notions related again to privacy and the maintenance of class values. An ardent critic of the picture window, John Keats called it "a vast and empty eye" that stared across the street at an identical aperture that reflected it and looked vacantly back again.[68] Even more troubling to Keats was his belief that the picture window was at the root of the loss of individuality, because, as he wrote, "In the American house, the picture eye in the tokonoma reflects the outside world; instead of representing the family, it represents other people's activities. It is specifically designed to turn attention outward, away from home."[69] Such a gross trespass against the development of inward-looking individualism had to be addressed, and Keats's voice was one among many to condemn glazing that exposed the family to outsiders or that incorrectly directed the family's view toward the street and neighbors.

Window walls and large amounts of glass also received criticism because they required constant maintenance. As Mary Catlin and George Catlin explained, sizeable windows were hard to keep clean, and "the servantless housewife is harassed and oppressed by a job which always seems to need to be done: getting at washing those pesky windows."[70] A dirty picture window could reflect very poorly on a housewife and her family, especially because dirt or lack of maintenance in any form signaled nonwhite and lower-class inhabitants.[71] Remembering also the classic admonition of hired, wage-earning housekeepers who "don't do windows!" it is easy to imagine that housewives would have associated window washing with work performed by hired laborers who were likewise generally of color and of the lower economic classes. As such, the picture window would certainly have provided a housekeeping nuisance, just as Mary Catlin cautioned. Indeed, the architectural and design literature from the period are filled with heated prose decrying the mind-numbing effects of the picture window on suburban inhabitants and linking it with crass, consumer culture.[72] But how were suburbanites to deal with this feature of their homes?

Far from providing an acceptable remedy, blinds and curtains, though almost always implemented in houses that contained large amounts of glass, were seen as a Band-Aid approach to solving privacy problems. After all, if one had to implement window coverings, why have the glass in the first place? Heavy window coverings and dark interiors also conjured prewar housing conditions and lower-class living. As a contributing author to *House Beautiful* wrote, "Unfortunately, in our best residential areas, obsolete restrictions created in times before the Glass Age prevent our putting fences, hedges, or walls close to our property lines and keep us from creating privacy, both indoors and outdoors. As a result, many people who responded to the urge for more sun and light are living behind drawn Venetian blinds and thin curtains to escape living like fish in a bowl."[73] To live in the "glass age" was to embrace the bright sparkle of the unimpeded view from the picture window and therefore to be among the modern, middle majority. But no one wanted to be so thoroughly on display,

If you are the only one who wants
privacy on your street,
here's one way to do it, beautifully

When privacy is not a common community goal, and you really want it, your strategy should be to create such landscaping beauty that all who see will admire and commend you. That you have achieved the by-product of privacy they may miss entirely while they marvel at your lush bower of beauty. This is the socially-acceptable way to achieve what you want without seeming to go against the local mores. The street-side planting explained above is a fine example of how it may be done. Two hedges, clipped differently, and a row of flowering trees were all it required. The low mounded hedge belies the true height of tall hedge. Luxuriant growth of all adds to attractiveness. Thomas Church, landscape architect.

Figure 7.6 The image at the top illustrates planting for privacy from the street. The middle section diagram illustrates a way to achieve privacy from the street and from neighbors. Reprinted by permission from *House Beautiful,* May 1960. The Hearst Corporation. All rights reserved.

to be exhibited like a household pet in its cage. Still, manufacturers and tastemakers cleverly marketed blinds to anxious consumers as "Windows that peeping Toms can't see through. But you'll be able to see from the inside out and enjoy the view from your picture window without feeling that you're on exhibition. No need for drawn shades, but the new Venetian blinds of clear plastic are so attractive you may want them anyway."[74]

Perhaps the only reconciliation attainable for the conflicts inherent in the picture window was the use of exterior privacy walls that shielded the window from passersby on the street but allowed a controlled view of a contained atrium garden to the family and permitted invited guests to peer indoors as they approached the house's front entry (see Figure 7.6). Still, such solutions seldom appeared in ordinary house lots, and the conflicts raised by the picture window raged in the press, as the battles over privacy and family life were waged within the home.

Conclusion

Overall, the concern with personal and family privacy achieved through manipulation of the design of house and garden connected to the desire for a private world, secluded from the contaminating and frightening influence of outsiders. Home and garden were best designed when they allowed one to turn one's back on the world. The images of domesticity displayed in the normative design literature of the period, with their texts that repeatedly emphasized the need for creation of private domestic realms, sold rapidly to readers whose anxieties over economic status, racial assignment, and claims to national identity they sought to assuage. Although they sold briskly, readers likely received the uniformly produced messages about domestic design variously, and information on reception remains obscure. Yet the design and construction of houses across the United States stand as ample evidence of the acceptance of the ideas promoted in the literature. Gated communities may be the latest and most extreme example of exclusionary design practices predicated on the desire for privacy, but the overwhelming exclusivity of suburban housing generally indicates the currency and significance such ideas have enjoyed. In the growing corpus of literature that examines the connections between the construction of race and that of the built environment, geographers and historians have examined the suburb and its planning in some detail. But as this chapter indicates, the house and the choices embodied in its design serve as indispensable markers of class distinction and racial identity.

Notes

1. In this essay, I follow the work of scholars in the field of critical studies of whiteness who sometimes place the descriptor "white" in quotes because whiteness is not an absolute description of a genetically determined racial group but instead a culturally constructed and determined category that is constantly in flux. On theories of whiteness, see for example, Theodore Allen, *The Invention of the White Race,* vol. 1 (London: Verso, 1994); Alastair Bonnett, "White Studies: The Problems and Projects of a New Research Agenda," *Theory, Culture, & Society* 13, no. 3 (1996): 145–55; Richard Dyer, *White* (London: Routledge, 1997); Peter Klochin, "Whiteness Studies: The New History of Race in America," *Journal of American History* 89, no. 1 (2002): 154–73; Aldon Nielson, *Reading Race: White American Poets and the Racial Discourse in the Twentieth Century* (Athens: University of Georgia Press, 1988); and David Roediger, *The Wages of Whiteness: Race and the Making of the American Working Class* (London: Verso, 1991). The field is expanding rapidly, but these sources are among those considered foundational. My thanks to Tim Engles and Dave Roediger for directing me to these sources. On the restriction of most new suburban homes to whites, see George Lipsitz, *The Possessive Investment in Whiteness: How White People Profit from Identity Politics* (Philadelphia: Temple University Press, 1998), 1–33. However, a growing body of literature also begins to identify suburban situations that were available to nonwhites. See for example, Andrew Wiese, *Places of Their Own: African American Suburbanization in the Twentieth Century* (Chicago: University of Chicago Press, 2004). Wiese demonstrated that the pace of suburban integration increased dramatically after about 1960, after the close of the era I examine in this chapter.
2. Joan Rosenbaum, "Foreword," in *Getting Comfortable in New York: The American Jewish Home, 1880–1950* (New York: Jewish Museum, 1990), 7.
3. On the cultural construction of privacy as it relates to domestic design, see Peter Ward, A History of Domestic Space: Privacy and the Canadian Home (Vancouver and Toronto: University of British Columbia Press, 1999).
4. The examples are far too numerous to list here, but for a few examples see *House Beautiful,* May 1960, 170; Elizabeth Mock, *If You Want to Build a House* (New York: Museum of Modern Art, 1946), 40; and Will Mulhorn, "The Enlarging Window—Straw-in-the-Wind..." *House Beautiful,* December 1946, 286.
5. See Ward, *A History of Domestic Space.*
6. The call for creation of a private residential world appears pervasively in the literature on domestic design in the postwar period, and I address selected examples in this chapter. For a sociological analysis, see Elaine Tyler May, *Homeward Bound: American Families in the Coldwar Era* (New York: Basic Books, 1988). Clifford Clark made a brief note of the desire for postwar domestic privacy in *The American Family Home, 1800–1960* (Chapel Hill: University of North Carolina Press, 1986), 219. As a planning and design

issue, privacy occurs countless times in articles in shelter magazines from the period.

7. David Sibley, *The Geography of Exclusion: Society and Difference in the West* (London and New York: Routledge, 1995), 78.
8. Ibid., 81, 94.
9. Kevin Fox Gotham, *Race, Real Estate, and Uneven Development: The Kansas City Experience, 1900–2000* (Albany: State University of New York Press, 2002), 126.
10. For Stuart Hall's contribution, see Hall, Chas Critcher, Tony Jefferson, John Clarke, and Brian Roberts, *Policing the Crisis: Mugging, the State, and Law and Order* (New York: Holmes and Meir, 1978), 394. See also Paul Gilroy, *The Black Atlantic: Modernity and Double Consciousness* (Cambridge, MA: Harvard University Press, 1993), 85.
11. For more on the racialized aesthetics of suburban spaces, see James S. Duncan and Nancy G. Duncan, *Landscapes of Privilege: The Politics of the Aesthetic in an American Suburb* (New York and London: Routledge, 2004).
12. Alan F. Westin, *Privacy and Freedom* (New York: Athaneum, 1967), 172, 174. Westin's chapter 8 is titled "Dissolving the Walls and Windows."
13. John Cheever, "The Enormous Radio," in *The Stories of John Cheever* (New York: Vintage Books, 2000), 33–41.
14. Westin, *Privacy and Freedom,* 179–81.
15. Vance Packard, *The Hidden Persuaders* (New York: David McKay, 1957), 42.
16. Westin, *Privacy and Freedom,* 279–80.
17. Westin, *Privacy and Freedom,* 281. Westin cited "Talk of the Town," *New Yorker,* September 21, 1957.
18. Nancy Duncan and James Duncan described an ideology of individualism rooted in private ownership, which they called a "basic tenet of bourgeois market society. Privately owned goods are markers of identity, even constitutive of identity." See Nancy G. Duncan and James S. Duncan, "Deep Suburban Irony: The Perils of Democracy in Westchester County, New York," in *Visions of Suburbia,* ed. Roger Silverstone (London and New York: Routledge, 1997), 164. Likewise, Joseph Howland, who was the garden editor of *House Beautiful* from 1948 to 1956, wrote that "for GIs whose parents raised their family as renters, the result was a spectacular interest in 'the face we present to the neighbors.' " See the biographical notes in the Joseph Howland Papers, Environmental Design Archive, University of California, Berkeley.
19. Winnie Breines noted that sociological texts of the 1950s were written by white male professionals and therefore represent a limited perspective. She noted, however, that Reisman's book was "extraordinarily successful" in its time. See Breines, *Young, White, and Miserable: Growing Up Female in the Fifties* (Boston: Beacon Press, 1992), 26, 27, 30. These texts, although about cultural production rather than records of reception, provide registers and indexes of cultural desire and anxiety, at least for a segment of the population.
20. Ibid., 9.
21. See ibid., 9–10. Breines wrote about "a policy of containment" that results from such fears, stating that "post war culture was a culture of containment."

22. Despite the Supreme Court rulings that declared restrictive covenants and zoning unconstitutional as early as 1917, the practices continued almost unabated for decades, implemented by real estate agents, political leaders, and bankers among others. Developers such as William Levitt, for example, refused to sell new postwar homes to those identified as nonwhite. Likewise, the Supreme Court's 1948 ruling in the *Shelley v. Kramer* case did little to end the enforcement of deed restrictions, because property owners could voluntarily enforce them, as could local authorities. As George Lipsitz noted, "After the Supreme Court decision in *Shelley v. Kramer,* the FHA persisted in its policy of recommending and even requiring restrictive covenants as a condition for receiving government-secured home loans." For a detailed discussion, see George Lipsitz, *The Possessive Investment in Whiteness: How White People Profit from Identity Politics* (Philadelphia: Temple University Press, 1998), chaps. 1 and 2, especially pp. 25–33.
23. David Reisman, *The Lonely Crowd* (New Haven, CT: Yale University Press, 1950), 306; and Lyman Bryson, *The Next America: Prophecy and Faith* (New York: Harper & Brothers, 1952). See especially chap. 12, "Individualism as Duty."
24. Reisman, *The Lonely Crowd,* 306.
25. Ibid., 307.
26. As the editors of *Fortune* magazine noted, "The modern suburbanite tries to keep down with the Joneses', or to put it more exactly, to consume no more and no less conspicuously than they. Not getting the balance just right is a source of friction, feuds, and sleepless nights in many of the newer suburban communities." See the editors of *Fortune, The Changing American Market* (Garden City, NY: Hanover House, 1953), 80.
27. Russel Lynes, *The Tastemakers* (New York: Grosset & Dunlap, 1954), 246.
28. Karal Ann Marling asserted that things seen advertised in the period, whether on television or in a magazine, were even more intriguing to 1950s home owners because once purchased, they existed "simultaneously in the public arena and in the private home," and she noted further, "An intense need for privacy was counterbalanced by a love of public display, introspection by extroversion, the primitive and simple by the technologically complex." See K.A. Marling, "Designing Popular Culture in the Postwar Era," Vital Forms: American Art and Design in the Atomic Age, 1940–1960, ed. B.K. Rapaport and K.L. Slayton (New York: Harry Abrams, 2001), 208–37. On the importance of house design to the cultivation of individuality, see also Sarah Williams Goldhagen and Réjean Legault, "Introduction: Critical Themes of Postwar Modernism," in *Anxious Modernisms: Experimentation in Postwar Architectural Culture* (Cambridge, MA: MIT Press, 2000), 15, 19.
29. Philip H. Ennis, "Leisure in the Suburbs: Research Prolegomenon," *The Suburban Community* (New York: Putnam's Sons, 1958), 262. See also John Seeley, *Crestwood Heights: A Study of the Culture of Suburban Life* (New York: Basic Books, 1956); John Keats, *The Crack in the Picture Window* (Boston:

Houghton Mifflin, 1956); and William H. Whyte, *The Organization Man* (New York: Doubleday, 1956).

30. Keats, *The Crack in the Picture Window,* 61, 169, 193.
31. Here I invoke the vocabulary popularized by the French sociologist Pierre Bourdieu. See in particular *Distinctions: A Social Critique of the Judgement of Taste,* trans. Richard Nice (Cambridge, MA: Harvard University Press, 1984).
32. Mock, *If You Want to Build a House,* 17.
33. Ibid., 93. Despite the popular notion that development houses were completely undifferentiated, recent studies indicate that home owners quickly personalized their residences. W.D. Wetherell made the point when his protagonist stated, "This little boxes stuff was pure phooey. Sure they were little boxes when we first started. But what did we do? The minute we got our mitts on them we started remodeling them, adding stuff, changing them around." See Wetherell, *The Man Who Loved Levittown* (Pittsburgh: University of Pittsburgh Press, 1985), 4. This point has now been verified in studies such as Barbara M. Kelley's *Expanding the American Dream: Building and Rebuilding Levittown* (Albany: State University of New York Press, 1993); and Rosalyn Baxandall and Elizabeth Ewen's *Picture Windows: How the Suburbs Happened* (New York: Basic Books, 2000). On the falsely apparent homogeneity of suburban houses, Simon Frith wrote, "What may seem homogenous from the outside—the rows of identical residences; the shared commuting lifestyle; a kind of cultural blankness—is marked from the inside (where the children live) by the recurring problem of fine difference." See Frith, "The Suburban Sensibility in British Rock and Pop," in *Visions of Suburbia,* ed. Roger Silverstone (London and New York: Routledge, 1997), 276.
34. See Matt Wrayle and Annalee Newitz, eds., *White Trash: Race and Class in America* (New York: Routledge, 1997). On "trailer trash," see also Andrew Hurley, *Diners, Bowling Alleys, and Trailer Parks: Chasing the American Dream in the Postwar Consumer Culture* (New York: Basic Books, 2001), 251–53.
35. Mary Catlin and George Catlin, *Building Your New House* (New York: Current Books, 1946), 162, 160.
36. Ibid., 163.
37. Kate Ellen Rogers, *The Modern House, U.S.A.: Its Design and Decoration* (New York: Harper & Row, 1962), 17, 21.
38. For more on Gordon, the Pace-Setter homes program, and the magazine during the twenty-year period of her editorship, see my essay "Making Your Private World: Modern Landscape Architecture and *House Beautiful,* 1945–60," in *The Architecture of Landscape, 1940–60,* ed. Marc Treib (Philadelphia: University of Pennsylvania Press, 2002), 180–205.
39. Elizabeth Gordon, "The Key to Pace-Setting Living," *House Beautiful,* November 1952, 212.
40. On "everyday modernism," see Marc Treib, *An Everyday Modernism: The Houses of William Wurster* (Berkeley: University of California Press, 1995).
41. Elizabeth Gordon, "The Responsibility of an Editor" (manuscript for speech delivered to the Press Club Luncheon of the American Furniture Mart,

Chicago, June 22, 1953), 14, 15, 21; Thomas Church Collection, Environmental Design Archives, University of California, Berkeley. My thanks to Marc Treib for bringing this document to my attention.

42. Joseph Howland, "Good Living Is NOT Public Living," *House Beautiful*, January 1950, 30.
43. On constructions of white and black sexuality, see Calvin Hernton, *Sex and Racism in America* (New York: Grove Press, 1966); and bell hooks, "Madonna: Plantation Mistress or Soul Sister," in *Black Looks: Race and Representation* (Boston: South End Press, 1992), 157–64.
44. See Ward, *A History of Domestic Space,* 137.
45. Ibid., 147. See also Virginia Scott Jenkins, *The Lawn: A History of an American Obsession* (Washington, D.C. and London: Smithsonian Institution Press, 1994).
46. "How Our Cars Have Changed Our Gardens," *House Beautiful,* November 1956, 254.
47. As a 1949 Pace-Setter house plan caption stated, "Broad L-shape living porch shows how vitamin-conscious moderns turn toward outdoors. Because people now work indoors, outdoors is a symbol of relaxation. Re barbecue: cooking has prestige when you cook for fun." *House Beautiful,* January 1949, 57.
48. Thomas Church, *House Beautiful,* August 1948, 78; Jean Burden, "From Ordinary Back Yard to Country-Club Living," *House Beautiful,* August 1958, 72; and "Mark of a MODERN House—The Paved Terrace," *House Beautiful,* November 1952, 220. Elaine Tyler May likewise noted, "The suburban home was planned as a self-contained universe ... family members would not need to go out for recreation or amusements, since they had swing sets, playrooms, and backyards with barbecues at home." See May, *Homeward Bound: American Families in the Cold War Era* (New York: Basic Books, 1988), 171; see also Clark, *American Family Home,* 219. Andrew Hurley argued that ultimately American families were pulled away from each other and into the outside world because "the vendors of mass goods and services recognized that men, women, and children were as likely to seek refuge from the straitjacket of family hierarchies and responsibilities through their consumption activities as they were to seek togetherness. ... The endeavor became more difficult as time wore on and the centrifugal pull of peer-group pressure and individual expression overwhelmed the centripetal lure of family togetherness." See Hurley, *Diners, Bowling Alleys, and Trailer Parks,* 290–91.
49. On the construction and currency of the familial myth in the 1950s, see Laura J. Miller, "Family Togetherness and the Suburban Ideal," *Sociological Forum* 10, no. 3 (1995): 394; Betty Friedan, *The Feminine Mystique* (New York: Norton, 1963), 48; and Andrew Hurley, *Diners, Bowling Alleys, and Trailer Parks,* 159.
50. A. Quincy Jones, *House & Home,* January 1957, 142. Although Jones here described his own custom-designed house, he was one of the architects (along with his partner Fred Emmons) who designed houses for the tract developer Joseph Eichler in California. Indeed, Jones maintained a strong

interest in designing and building low-cost housing for the middle-majority buyer, a feature that remains a distinguishing hallmark of his career.

51. "The Backyard—America's Most Mis-Used Natural Resource," *House Beautiful,* January 1949, 37–42. See also Environmental Design Archives, University of California, Berkeley, Folder: Garden Design Plants-Alive Environment.
52. Elizabeth Gordon, "Is Privacy Your Right or a Stolen Pleasure?" *House Beautiful,* May 1960, 152, 232, 234–35.
53. See for example the case of the Myers family, whose purchase of a home in the Pennsylvania Levittown sparked a race riot and two months of intense harassment. On that case, see Marvin Bressler, "The Myers' Case: An Instance of Successful Racial Invasion," *Social Problems* 9 (Fall 1960): 126–42. See also Andrew Wiese, *Places of Their Own,* 156; and Mark Clapson, *Suburban Century: Social Change and Urban Growth in England and the United States* (Oxford and New York: Berg, 2004), chap. 4.
54. Karen Brodkin, *How Jews Became White Folks and What That Says about Race in America* (Piscataway, NJ: Rutgers University Press, 2000).
55. Douglas Baylis, "Privacy Makes Better Outdoor Living," *Better Homes and Gardens,* April 1950, 48–54. See also Douglas Baylis Collection [1999-4], Office Records/Clippings, Environmental Design Archives, University of California, Berkeley.
56. The X-100 received a great deal of media attention in the year of its construction, but the privacy wall was featured in *Concrete Masonry Review,* April 1957, 16–17.
57. "Seclusion by Design," *House and Garden,* January 1957, 34–35.
58. Catlin and Catlin, *Building Your New House,* 149.
59. According to Marc Treib, the landscape architect Fletcher Steele recommended as early as 1924 that kitchens be moved to the street side of the lot to facilitate a greater connection between living and dining spaces and the garden. Indeed, this pattern is found in postwar housing as well, but equally numerous are houses that moved the kitchen toward the rear of the house or allowed it to span the length, from front to back, in very small houses. See Marc Treib, "Aspects of Regionality and the Modern(ist) Garden in California," in *Regional Garden Design in the United States,* ed. Therese O'Malley and Marc Treib (Washington, D.C.: Dumbarton Oaks, 1995), 15. For an interesting analysis of actual use versus architect's intentions in an Eichler house, see Annmarie Adams, "The Eichler Home: Intention and Experience in Postwar Suburbia," in *Gender, Class, and Shelter: Perspectives in Vernacular Architecture,* ed. Elizabeth Collins Cromley and Carter L. Hudgins (Knoxville: University of Tennessee Press, 1995), 164–78. Adams showed that despite the designer's ideas about supervision from the windows, children managed to find ways to escape their mother's watchful eye.
60. Elizabeth Gordon, "The 12 Best Houses of the Last 12 Years," *House Beautiful,* September 1947, 89.
61. Carolyn Murray, "How an Interior Court Can 'Save' a Crowded Lot," *House Beautiful,* November 1957, 305. See also "Orientation Is More Than Just a

Word," *House Beautiful,* May 1961, 109; and Madelaine Thatcher, "How Public Should the Front Entrance Be?" *House Beautiful,* November 1963, 225.

62. George Marcus claimed, "William Levitt's rules excluded fences and walls," yet Levittown became a suburb with fences eventually. See George H. Marcus, *Design in the Fifties: When Everyone Went Modern* (Munich and New York: Prestel, 1998), 40.
63. Thomas Church, *Your Private World: A Study of Intimate Gardens* (San Francisco: Chronicle Books, 1969).
64. On Eichler homes, see Paul Adamson and Marty Arbunich, *Eichler: Modernism Rebuilds the American Dream* (Salt Lake City, UT: Gibbs Smith, 2002).
65. Although historians such as Lynn Spigel have correctly asserted that picture windows served as essential showcases in which to display, they have largely neglected to address the privacy problems that arose in conflict with such desires. See Lynn Spigel, "From Theatre to Space Ship: Metaphors of Suburban Domesticity in Postwar America," in *Visions of Suburbia,* ed. Roger Silverstone (London and New York: Routledge, 1997), 221.
66. Sandy Isenstadt, "The Rise and Fall of the Picture Window," *Harvard Design Magazine,* Fall 1998, 30.
67. Sandy Isenstadt, "The Visual Commodification of Landscape in the Real Estate Appraisal Industry, 1900–1992," *Business and Economic History* 28, no. 2 (1999): 65.
68. Keats, *The Crack in the Picture Window,* 21, 58–59.
69. Ibid., 167. *Tokonoma* is the term used for a decorative alcove typically found in the main hall of a Japanese noble residence. Keats's use of it here is unusual, but he seems to use it to satirically portray the picture window as the sacred alter of the suburban house in the United States.
70. Catlin and Catlin, *Building Your New House,* 150.
71. Suellen Hoy, *Chasing Dirt: The American Pursuit of Cleanliness* (New York: Oxford University Press, 1995), chap. 6.
72. See Isenstadt, "The Rise and Fall of the Picture Window," 30–33.
73. Will Mulhorn, "The Enlarging Window—Straw-In-The-Wind That Shows You the Direction in Which American House Design Is Heading," *House Beautiful,* December 1946, 286.
74. Judy Barry, "Report: To the Housewife," *Ladies' Home Journal,* November 1945, 195.

CHAPTER 8

Aesthetics, Abjection, and White Privilege in Suburban New York

JAMES DUNCAN AND NANCY DUNCAN

Introduction

Cultural landscapes can evoke powerful images and emotions that help to constitute a moral order and can play a central role in the practices and performance of place-based social identities, community values, and social distinction (Cosgrove 1993; Daniels 1993; Graham 1994; Lowenthal 1991; Rose 1995; Matless 1998). This is especially the case when the landscapes in question also are landscapes of home. Landscapes of home are personal, even as they link us to a place-based sense of community. The landscape of home often is a commonly recognized landscape that plays an active role in articulating identity, differentiating social position, and maintaining both positive and negative emotional attachments to place. Landscapes of home thus can be fraught with politics. They are implicated in the work of defining and negotiating just who belongs to a community (and how they may belong). Powerful localized interests often are able to define a seemingly stable, rooted, and closed local community through the cultural landscape, despite the inevitable interconnectivity and mutual interdependence of the "local" and the "global" in contemporary life. This is especially the case in affluent U.S. suburbs.

In this chapter we explore the racialized politics of landscapes of home and identity in two adjoining suburban towns in Westchester County near New York City. These suburban landscapes result from conscious design practice and political struggle, and we argue that because of the cultural emphasis in these landscapes on the aesthetic and the visual, the suburban landscape can be said to aestheticize and obscure social and cultural antagonism and unacknowledged racism. We based this chapter on material drawn from our ethnographic monograph based on three decades of research, including more than two hundred semistructured, in-depth interviews, three citizen surveys, attendance at local planning board and NIMBY (not-in-my-backyard) meetings, as well as analysis of newspaper reports, official publications, and local archival materials.

The two towns are Bedford and Mount Kisco. Bedford appears as a rural idyll that is nevertheless reliant on a set of (largely obscured) globalized and racialized social relations. Bedford's landscape aesthetic implicates its proponents in the practices of white privilege that link it to the adjacent suburban landscape of Mount Kisco, especially through its labor force needs. Mount Kisco is a small commercial hub serving Bedford and several other rural villages struggling with a perceived Latino invasion that has resulted in vociferous debates about loitering and the proper aesthetic of suburban public space. It is to these particular landscapes that we turn next, even as we suggest that the practices and performances of a suburban aesthetic in these places point to similar issues for suburban landscapes nationwide.

Bedford and Mount Kisco: Suburban Aesthetics

Bedford Village is a picture-perfect settlement of white clapboard colonial houses and stores with a historic village green surrounded by wooded and pastoral landscapes. Once a community of farmers, Bedford later became a suburb of late nineteenth-century and early twentieth-century gentleman's estates that by the 1960s had acquired a slightly seedy elegance with a distinctly English pastoral flavor. Today residents are spending considerable sums on landscaping; restoring historic mansions, barns, and stone walls; and generally maintaining Bedford's pastoral beauty. The town has more than a hundred miles of dirt roads, has horses grazing on its many open fields, and has forests of spruce and hemlock. Its rural landscape aesthetic is supported by a zoning requirement that applies to 80 percent of the land area, which states that each house have at least four acres of land. Conservation easements, strict enforcement of stringent environmental legislation, and a proliferation of small nature preserves and wildlife corridors further protect many of Bedford's largest and most beautiful properties from being developed, even at the legal four-acre lot minimum.

Although considerably smaller in area, Mount Kisco is much more densely developed than Bedford. Nevertheless, it has a pleasing, small-town Norman Rockwell streetscape of Victorian and early twentieth-century storefronts. It is composed of a few commercial streets and various small residential districts catering to a variety of income levels, ranging from low-income subsidized housing to large Victorians occupied by middle-class to upper-middle-class families. Although most streets in town are lined with maples and other large trees, there are a few rural roads with long driveways leading to some large estates. Until recently many Mount Kisco residents saw it as an all-American town made up of successful, long assimilated Italian American and other ethnic groups, with a small section of black families with well-established roots in the town. It was once described as "a heterogeneous blue collar community, a medical, automotive and shopping center"; "slightly more than two square miles replete with Elks, Lions, Masons, Rotarians, Kiwanis and an active chamber of commerce." Although the village retains these elements, it has recently attracted some younger, highly educated residents who would like to help point Mount Kisco in a more gentrified direction and a large Latino community of poor laborers who have become the source of much discontent in this otherwise seemingly contented town.

Place-Based Identity

Collective memories, narratives of community, invented traditions, and shared environmental ideals and fears are repeated, performed, contested, and stabilized in the landscape's material form. Various social, economic, political, and legal practices have evolved to create and fix the association between such ideas and landscapes. These practices tend to be exclusionary although they are not always acknowledged, or even recognized, as such. Usually they are defined as preservation. And in fact the goal is not always social exclusion in itself but to preserve the "look of the landscape," which is central to the performance of particular social identities that depend on lifestyle, consumption patterns, taste, and aesthetic sensibilities.[1]

Members of small, affluent, and relatively homogeneous communities such as Bedford (and to a somewhat lesser extent, Mount Kisco) can mobilize sufficient economic and cultural capital to preserve or recreate landscapes that have the power to establish, incorporate, and assimilate some identities while excluding or erasing others. They are aided in this by the very high degree of local autonomy allowed by the structure of the American political system. These landscapes become scarce positional goods charged with an aura of distinctiveness and authenticity that can lend prestige to those associated with that place. In the United States, where identity is linked to possessions (including houses and land, which are in turn valued by the prestige and attractiveness of their location), the aesthetic can play a significant role in depoliticizing the privileges of class and race as constituted by power, authority, and production practices (Harvey 1989). By this we mean that social relations are obscured through the aesthetic and are incorporated into categories of lifestyle, taste, consumption patterns, and appreciation of the visual, the sensual, and the unique. Although some geographers and sociologists have long understood that landscape taste is an important positional good (Duncan 1973, 1999; Duncan and Duncan, 2004; Firey 1945; Higley 1995; Hugill 1986, 1989; Ley 1993, 1995; Lowenthal 1991; Lowenthal and Prince 1965; Pratt 1981; Wyckoff 1990), we argue that it is more important as a form of cultural capital (Bourdieu 1984) than many academics have recognized. Landscapes, especially those with pleasing views, become possessions for those with the wealth and power to control them; thus aesthetic appreciation of residential landscapes is an issue that preoccupies primarily the affluent. Perhaps as a consequence, landscape taste and the aesthetic aspects of social relations in place have been considered relatively inconsequential and are rarely investigated by academics. However, as we argue, these have unseen social consequences, more far reaching than may at first appear. A seemingly innocent pleasure in the aesthetic appreciation of landscapes and efforts to maintain and protect them can act as subtle but highly effective excluding mechanisms for reaffirming class and race identities.

Bedford Village: "Can't Live with Them, Can't Landscape without Them"[2]

Bedford Village has large estates, small farms with rolling hills, and horses grazing on open meadowland, as well as suburban colonial-style houses on four-acre lots as required by the town zoning code. Bedford's beauty lies in the rurality of its landscape. Such a landscape including miles of dirt roads, stone walls, and bridle paths requires labor-intensive care and maintenance, which is provided today by recently immigrated Latino day laborers. It also requires highly sophisticated political organization to exclude undesired development. The aesthetic value of having a sparsely populated, rural landscape is seen by virtually all of the town's residents as unquestionable.[3] According to an aestheticized view of nature, Bedford's many acres of pasture, forests, and large wooded house lots indicate that it has more "nature" than if the town were composed principally of houses surrounded by manicured lawns. This romantic discourse lends support to the exclusionary structures and practices maintaining Bedford's scenic landscapes. We believe that the celebration of the natural environment, the historic preservation, and the professed uniqueness of this landscape divert attention away from the interrelatedness of issues of aesthetics and identity on one hand and social justice in the form of housing opportunities and related advantages such as good schools on the other.

This potential tension between the aesthetic and social justice presents a political problematic for understanding and analyzing landscapes of home. The aesthetic is an embodied, affective sensibility not always clearly articulated. That is, the aesthetic is taken as natural or normal. One can present an intellectual argument for the political implications of taking pleasure in landscapes (Rose 1993). It also is possible to understand historically how Western politics surrounding the environment and ethnonationalist politics can be aestheticized if they are linked to romanticism, including aesthetic appreciation and contemplation of the picturesque, the beautiful, and the sublime in nature and a deep emotional and sensual attachment to land. Nevertheless, this history and politics often is unfamiliar to those immersed in local political struggles over landscape appearance. Questions about aesthetic appreciation are generally seen as personal, spontaneous, and nonideological—a matter of personal taste that happens to be shared by like-minded, similarly educated members of one's community. Although the aesthetic is seen as vulnerable to politics, it is not usually thought to be ideological or political in and of itself.

The desire to protect nature and history and the seemingly innocent pleasure derived from natural and historic landscapes can be understood as a process of displacement whereby moral–political issues are forgotten as attention is focused on aesthetic concerns.[4] This raises the difficult issue of complicity whereby individuals' actions and lifestyles are implicated in harm done to others through their association with many other people and institutions with similar goals. As Kutz (2000, 1) said,

> We find ourselves connected to harms and wrongs, albeit by relations that fall outside the paradigm of individual, intentional wrongdoing. … [Although] we stand outside the shadow of evil, we still do not find the full light of the good. Even individual acts … are characterized by a whole spectrum of relations between agents and harms, doers and deeds.

This view of complicity challenges the deeply engrained modern individualistic conception of moral agency that exempts individuals from being held accountable for the consequences of joint action where their own contribution is minor (Kutz 2000). We believe that certain unquestioned "goods" such as environmental conservation and historic preservation can have unintended harmful consequences for which individuals may not be accountable qua individuals but be seen as complicit. People in Bedford speak of landscapes symbolizing, and even inculcating, political and moral values, as well as conveying social distinction. Self-assured in the righteousness of their attempts to maintain open green space, the residents' aesthetic pleasures can be sustained only through spatial separation. Residents largely succeed in spatially and visually insulating themselves from uncomfortable questions of race and poverty by keeping out of sight reminders of the social consequences of their privilege. This privilege is what Pile (1994) referred to as "painless privilege" or what Pulido (2000, 13) referred to as "white privilege," which she defined as an unconscious form of racism resulting from a lifelong inculcation (Devine 1989; Lawrence 1987) that takes as natural "the privileges and benefits that accrue to white people by virtue of their whiteness."[5] They know that their landscapes depend on a politics of antidevelopment and the barricades against outside negative externalities. Nevertheless, although remaining aware of this, they tend to naturalize their privilege, having no reason or desire to trace the far-reaching, unintended consequences and complex conditions of their privilege.

Bedford is highly interconnected within transnational flows and networks of power, privilege, and economically and politically driven migration mainly from Latin America. Many CEOs of large corporations live in Bedford, whereas regional- and national-level political organizations,— such as the American Civil Liberties Union, the Westchester Hispanic Coalition, and the Center for Immigrant Rights all based in New York City as well as the Nature Conservancy and Westchester Land Trust—are examples of large-scale institutions that have been enrolled in networks that connect northern Westchester to the wider region, the nation, and beyond. It is clear that as institutions, such as Spanish-language newspapers and transnational labor, migrant, and other political organizations, are spreading out of New York City, the suburbs are becoming more highly connected and, from the point of view of many residents, increasingly exposed to unwelcome "infiltration" and penetration by heterogeneous outside forces. The demand for gardeners and cleaners, restaurant employees, and various unskilled and semiskilled laborers, as well as low-waged clerks and casual flexible labor is bringing new immigrants into northern Westchester, especially Latinos. The subsequent pressure on the lowest end of the housing market has resulted in their exploitation by the owners of the relatively small number of rental housing units as well as some cases of homelessness. What are seen as urban people and problems coming into the suburbs are distressing to longer-term residents used to enjoying their painless, white privilege without visual reminders of poverty.

Abjection, Anxiety, and the Latino Presence in Mount Kisco

The visible imperatives of globalization are especially evident in Mount Kisco, where many of the day laborers who maintain Bedford's landscape live. Mount Kisco struggles with a politics of the aesthetic that is intimately linked to the rural aesthetic in Bedford and is openly inflected by race. The burning political issue in the Village of Mount Kisco during the past decade has been a perceived invasion of Hispanics[6] or "Guatemalans," as they are often called locally by non-Latinos.[7] Poor Latino immigrants living in Mount Kisco are part of a growing trend in the United States. They constitute what Mike Davis called "a far reaching 'Latin Americanization' " of U.S. metropolitan areas, including New York and its suburbs. According to the 2000 U.S. Census, Latinos constitute 25 percent of Mount Kisco's population, but this figure is a minimal estimate. Latino immigrants have been drawn to places like Westchester County because they are able to fill a growing niche in the local service economy. Many of these are seasonal, nonunion jobs with few benefits and little security (Zavella 2000). We believe that the perception of an invasion can be explained in large part by conflicting cultural conventions of public space based on an ethnocentric and class-based aesthetic. In fact, we argue that the day laborers are a cause for anxiety, resentment, and aversion for aesthetic reasons more than economic reasons, as they tend not to compete for jobs with any other group in the area. This is certainly not to deny that deep structural, economic, and political inequalities explain why the Central Americans find themselves in such a hostile North American environment or to deny that their poverty contributes to the racialization of many Latin Americans. Rather, it is to say that economic factors alone are insufficient as an explanation of the deep-seated psychological insecurities that shape social relations between immigrants and nonimmigrants in Mount Kisco.

The immigrants who stand in groups on the street corners of Mount Kisco chatting or waiting for contractors are there in large part because of the increasing demand for their labor on Bedford's large estates and smaller would-be estates. They help to sustain the narrative structure of Bedford's landscape by recreating and maintaining its walls, gardens, lawns, and country houses. And yet Bedford residents see their presence even in Mount Kisco as a very mixed blessing. Those who provide their labor to maintain Bedford's landscape aesthetic are considered an unaesthetic element of the streetscape in Mount Kisco, where Bedford residents go for shopping and services. And many residents of Mount Kisco and other nearby towns resent their towns becoming what they describe as "dumping grounds" or "servant's quarters" for places such as Bedford.

The Latino day laborers' physical appearance, deportment, and ways of using public space are subject to strong feelings of aversion (Sibley 1998; Atkinson and Laurier 1998; Philo 1998), making them the intended objects of extraordinary exclusionary measures.[8] As we explain, non-Latinos who live in Mount Kisco have had more difficulty in maintaining a homogeneous landscape of home that enhances their desired performance of successful middle-class American identities. Consequently they have had to confront their own attitudes toward the presence of racialized ethnic and class "others" who share their residential space. Many non-Latinos find this deeply threatening not only to the performance and representation of their desired identities but also to their self-perceptions as tolerant and nonracist. The illusion of a stable homogeneous place is ruptured by the very obvious visual presence of newly arrived outsiders and can lead to profound anxiety and aversive racism that perceives the other as abject. It is a kind of fearful loathing that results when a sufficient number of visually unassimilable outsiders threaten the physical and psychological borders of the self and thus undermine identity. The abject Latino presence upsets the proper functioning of an order; in the case of the Mount Kisco townscape it is a visual, spatial, and moral order. Although the objects of abjection may not always be considered intrinsically polluting or dangerous, they are seen to become so when deemed out of place and uncontrolled.

Different people view the Latinos as abject for slightly different reasons. For less affluent non-Latinos of Mount Kisco, it may be because their place-based identity is undermined by living spatially and economically close to members of a disdained group. On the other hand, some residents of both Bedford and Mount Kisco may feel a more aversive, unacknowledged[9] racism because they reject discursive racism and thus cannot admit even to themselves that their reactions should be labeled racist. The problem, Young (1990) said, is that aversion to people of another race, though profound, is largely unacknowledged as racism. Such reactions are primarily bodily, material, and unconscious. They include nervousness, avoidance, disgust, and distancing.[10] Local examples of aversive racism are negative aesthetic reactions to the physical presence of Latinos in Mount Kisco, to the sounds of their speech and music, and to their consumption styles.

Abjection in this context also is unsettling for its ambiguity. Kristeva's (1982) concept of the abject claims that ambiguity is threatening or loathsome, adding to the visceral and psychological dimensions of our understanding of aversive racism. In the East Coast American context, where race has largely been defined in terms of a black–white dichotomy, brown-skinned Latinos who are of Indian and mestizo background represent an uncomfortably ambiguous racial category for the non-Latino residents. These ill-defined, porous boundaries lie at the heart of abjection. When one's social relations with others are ambiguous or indeed if they seem uncontrolled, illegal, or disrespectful of norms, then these relations become difficult to tolerate. Local opinion in Mount Kisco ranges from what might be termed moral panic[11] of extreme nativist reaction[12] to paternalistic "tolerance," including a desire to help the newly immigrated assimilate to "proper American middle-class" ways of behaving in public space. Most of the remarks about Latinos made in the public opinion survey reveal an implicit, unreflexive form of nativism.[13] Among many local non-Latinos there is a strong, visceral distaste for the foreign-looking, indigenous American features and skin color of poor Latino men whom they accuse of loitering.

Loitering here is racialized, gendered, and class based. The difference in appearance and ways of being male of those socializing or waiting for work on street corners defines them as loiterers. That difference makes their behavior offensive and threatening to middle-class and upper-middle-class non-Latinos.[14] The same behavior on the part of unmarked middle-class whites, including middle-class, light-skinned Latinos, might not be so immediately defined as loitering.[15] Edward Chacon, a Guatemalan immigrant who in thirteen years progressed in Mount Kisco from dishwasher to landscaper to financial analyst at Citibank, provided an example of the way race is crosscut by class.[16] He found that acceptance is largely based on appearance. "When I wear my name tag from work in the street, I'm treated with respect. If I dress casual, they treat me like garbage" (Gross 2000a). Graciela Heyman of the Westchester Hispanic Coalition also argued that discrimination is not just about race: "It's a class thing. Because the men are skuzzy, they're short and they're brown" (Gross 2000a).

Identity in the United States, on the part of nonimmigrants, is often defined *against* and in contrast to an outside world beyond its borders. To many, the very bodies of undocumented immigrants can be said to act as a metaphor for insecure national boundaries. As Mains (2000, 151) said, immigrant bodies are marked as "separate, marginal, [and] different." The idea of insecure borders is aggravating or alarming to many nonimmigrants (Price 2000). Furthermore, many residents feel that the visibility of racialized difference, the phenotypic differences, act as daily reminders of the vulnerability of Mount Kisco to the negative externalities of more affluent towns: "Our town is a dumping ground." "We want our town back!" people said in the village survey. Furthermore, in the United States the visible presence of males congregating in public spaces is often seen as signifying a challenge to the dominant public order, which is highly individualistic and favors private spaces for socializing. Such behavior is commonly associated with gangs, political protests, and general male unruliness. Nevertheless, nonimmigrant attitudes toward immigration are ambivalent (Nevins 2002, 95–122). The historical narrative of the successful assimilation of immigrants into a vast melting pot is widely celebrated. Contemporary immigration, however, is often viewed with suspicion. Even when it is welcomed or tolerated, there remains a widespread NIMBYism due to concern over property values, even if for no other reason. Thus some are driven to become complicit with racism against their better intentions.

The poorer, less well-educated groups in Mount Kisco are sometimes quite explicit in their racism. Ironically, this includes some of the children, grandchildren, and great-grandchildren of Italian immigrant laborers who came to Westchester to build the late nineteenth-century and early twentieth-century walled mansions with gated entrances and the gardens, many of which are now being restored by Latino laborers. Many of these Italian Americans, whose ancestors suffered hatred and spatial exclusion in the late nineteenth century (when Italian immigrants rioted in Mount Kisco protesting against unfair working conditions, and the whole village of Katonah was deed restricted against selling or renting property to Italian immigrants), are as quick as any one else to urge the village to "solve" the "Hispanic problem."

Although some residents may indeed enjoy the color and enrichment of ethnic restaurants, world music, foreign travel, and multiculturalism in their urban experiences, they choose a more familiar, culturally homogeneous, safer aesthetic at home precisely because their liberal ideology requires them to embrace a narrative of equality that disallows overt racism.[17] Because of affluent Americans' aesthetic mode of treating foreign subjects, setting becomes important. Foreigners appreciated as romantic and colorful in their "proper" foreign place can become a repugnant and intrusive presence in American home spaces where a secure and stable retreat from the challenges of a globalizing world is sought. Many such people expect their home space to provide continuity with either real or imagined landscapes of childhood—to provide them with a mirror of their social selves and perhaps, more important, their memories or fantasies of the good life. To achieve this, they must have their home space "purified."

To Young's notions of dominative and aversive racism can be added "white privilege," which "thrives in highly racialized societies that espouse racial equality, but in which whites will not tolerate either being inconvenienced in order to achieve racial equality—or denied the full benefits of their whiteness" (Pulido 2000, 15). White privilege is a powerful force, Pulido argued, precisely because most whites are unconscious of it and thus they can exonerate themselves from racism.[18] She (Pulido 2000, 16) argued, "The full exploitation of white privilege requires the production of places with a very high proportion of white people. 'Too many' people of color might reduce a neighborhood's status, property value, or general level of comfort for white people." Pulido said that white privilege underlies institutional racism. We add that, as well as underlying institutional racism, white privilege is enabled by the institutionalized exclusion that results in the relative homogeneity of towns like Bedford. Residents need not be racists, or at least not confront any racism they may harbor, to enjoy their privilege because of institutional racism such as the Federal Housing Authority's mobilization of bias against nonwhite home owners.

Mount Kisco has somewhat more rental housing than Bedford, which has very little. However, the Latino day workers' housing situation is generally appalling. Local landlords can exploit Latinos because there are few low-rent apartments available anywhere in northern Westchester. Landlords illegally rent basements, attics, and rooms to laborers, as many as ten to a room sleeping in shifts. Although the village has tried various measures to crack down on such poor housing conditions, the motivation behind these measures is clearly to drive Latinos out of town, as there have been no attempts to revise the village zoning code to allow for more rental housing. In fact the trend has been in the opposite direction—to further strengthen the exclusionary structures that underlie, and tend to remain, the invisible forces in this story.[19]

Many of Mount Kisco's Central Americans have inadequate private space in which to entertain their friends, and furthermore they have a culture of socializing in large groups in public; this is especially true of the men.[20] However, the presence of racially marked outsiders offends the aesthetic of homogeneity necessary to maintain the appearance of community. It is not so much the actuality as the visibility of poor Latinos in the area that disrupts the spatial, moral, and visual order of white suburban society. Many of our informants put it bluntly: the very presence of the Latino day laborers on village streets is thought to "spoil the look" of the landscape. Such concerns are far from unique to Mount Kisco (Belluck 2005; Berger 1993; Carvajal 1993; Cooper 1999; Gross 2000a, 2000b; Mitchell 1997; Ortiz 2000; Toma 2001).

Even though the village public opinion survey did not specifically ask about loitering, a large number of the respondents (118) offered optional written comments, citing loitering as the major problem facing Mount Kisco.[21] In response after response, residents mentioned loitering or hanging out, and in others loitering is alluded to through the use of code words such as "quality of life." In response to the question about what detracts from the quality of life in Mount Kisco, a large number of answers referred to loitering, for example: "Groups of day laborers grouping downtown and sitting around the town on park benches." Examples of other responses mentioning loitering are as follows: "loitering detracts from the quality of life in Mount Kisco," "the men that are illegal and stand around Main Street," and "too many illegal people standing around." The term "illegal people" spoken thoughtlessly is revealing of the attitude that the Latinos of Mount Kisco have come to embody their immigration status. The following quotation gives a sense of how strongly many people feel about loitering: "Loitering takes away from the village. Many people will not come to town." On many of the questionnaires the issue of loitering is slipped in wherever possible in answer to various types of questions. The naturalized concept of loitering is culturally and historically specific in ways often unacknowledged by those who complain. In Latin America, far more than in the United States, civic and social life takes place in plazas and other public places. Latinos, on moving to the United States, are expected to abandon their normal modes of interacting and ways of making space attractive and inviting. Often this expectation is communicated to Latinos as a "helpful" part of what non-Latinos see as a necessary assimilation process.

One can see in the following remarks that loitering is often seen in aesthetic terms as visually offensive behavior: "the loitering problem on Maple Avenue and along Lexington is getting to be an *eyesore,*" "the *beautiful* benches are occupied by day workers," "young men perched on the benches" (are) "*unsightly,*" and "illegal aliens [are] *cluttering up* the Kirby Plaza." Given the large number of comments made about noise and the sounds of Spanish being spoken in the village, we can only assume that people object to what they see as aural as well as visual pollution. "The new immigrants take away from the beauty of the village by hanging around and not learning the language." Others remarked that they "despised the Latinos who hang out in the town on the street corners and gazebo." They argued, "Day laborers shouldn't be in Kirby Plaza." Presumably such comments are made unreflexively without irony and with little thought about what plazas, gazebos, and benches are designed for or what the definition and cultural history of the *plaza* are (Low 2000).

Loitering is sometimes seen as threatening. The fact that Hispanic men make comments to women and are sometimes perceived as "leering" and "eyeballing women" adds to the intimidating appearance of the day workers. Residents made comments on the questionnaire such as "can't walk around town at certain times without comments from people hanging out on the streets," "loitering detracts—way too many men roaming, too many groups hanging around the town is very intimidating," "the look is disturbing sometimes for a single woman," "there are too many people hanging out on the village streets, day workers without proper police presence," "there are too many Hispanics hanging out on the street corners and a lack of police presence," and "loitering of rough types at Kirby Plaza and the train station."[22] One respondent specifically mentioned that Kirby Plaza in front of the village train station was built in memory of one of Bedford's leading citizens and that it was an affront to his memory that the plaza has become "overrun with Latino males." The thrust of these comments about loitering in the village survey is that the quality of life of residents would be improved if the day laborers were forbidden from congregating in public spaces to look for work or meet with friends. The notion of quality of life, in this sense, as Mitchell (1997, 326) pointed out, prioritizes the aesthetic values of the middle class over the survival of the poor.

The solution to erasing the Latinos from the streetscape of Mount Kisco, which took many years to negotiate,[23] entailed setting up a hiring center in a warehouse district in the most marginal, out of the way location in the village possible. Our most recent interviews suggest that there may be a significant reduction in the degree of moral panic, as the day laborer hiring sites are no longer located on the street corners in the center of town and the facial features and body types of the laborers who do still stand around in the center are becoming somewhat more familiar. The issue of growing tolerance, however, does not negate aversive racism, for as Mendus (2000) pointed out the concept of tolerance suggests that there is something abhorrent to be tolerated, not so abhorrent that it cannot be tolerated but abhorrent nevertheless.

Conclusion

Place-based identities are maintained in large part by excluding or making invisible that which is incompatible with maintaining a stable sense of self and community. Because American space is conceived of as divided up into discrete home places that are central to people's identities, residents seek the visual appearance and ambience of congeniality and familiarity, even when the actual fact of homogeneity is not achievable. Attachment to a landscape of home has bodily, visceral, and affective components that are defensive and exclusionary but tend to remain relatively unarticulated. This may produce unacknowledged racism that sits uncomfortably with more liberal ideals and the celebration of cultural difference in the abstract. Foreign-looking bodies, especially when they are new in a place, can disrupt the illusion of community and its historical roots. It becomes increasingly difficult to maintain the fiction of a collective "we" as in "our forefathers who founded this town" or "our colonial heritage" (see Duncan and Duncan 2004, chap. 7).

People in Bedford might not realize it, but they are nevertheless complicit in the (unintended) consequences of the execution of their landscape ideal. The landscape they take for granted as their privilege may appear natural, but it is reliant on an increasingly globalized and racialized set of social relations. In Mount Kisco, on the other hand, the physical presence of Latino immigrants in public space acting outside of their caretaking role create aversion and anxiety, serving as unwelcome reminders of the globalized social relations that remain invisible in Bedford.

Places like Bedford and Mount Kisco have yet to make room for a reorientation of the spatial imagination such that place might be conceived of as the meeting up of histories or a multiplicity of trajectories (Massey 2005). To supplement theoretical attempts to think about space differently and more progressively (Agnew 2005; Massey 1991, 1993, 2005), we think it is necessary to further investigate empirically the noncognitive, emotional, and visceral reactions to place as well as the importance of aesthetics and illusion to the performance of place-based identities. We believe that to understand not only the taken-for-granted privilege but also the individualism that underlies refusals to acknowledge complicity in collective harms, we need to further explore ethnographically the lived, everyday difficulties that have to be overcome to move toward new, more open, and nonracist and indeed nonracialized ways of imagining place.

Notes

1. In Duncan and Duncan (2004) we elaborate the cultural background and politics of this orientation toward aesthetics.
2. Taken from Purdy (2001).
3. See Duncan and Duncan (2004), especially chapter 4. The citizen surveys conducted by the town and our own interviews overwhelmingly confirm this.
4. Normally to take an aesthetic attitude toward something is to react to it sensually not analytically—not looking beneath its surface to study or criticize the underlying social relations and other conditions of its production or reproduction (see Harvey 1996, 8).
5. On white privilege, see also Dyer (1997), Frankenberg (1993), and McIntosh (1988).
6. We use the self-ascribed terms *Latino* or *Latina* except in quotations and where the official term *Hispanic* is normally used.
7. For a striking parallel on Long Island, see Baxandall and Ewen (2000, chap. 17).
8. See Duncan and Duncan (2004, chap. 8) for an account of such exclusionary measures and forms of resistance to these measures, including court battles.

9. Etienne Balibar (1991, 40) pointed out, "There is not merely a *single* invariant racism, but a number of *racisms*."
10. Iris Marion Young (1990, 141–42), following Joel Kovel (1984), distinguished between dominative and aversive racism. The former she described as an openly admitted and practiced racism, whereas the latter is a racism of avoidance and separation. Although dominative racism has characterized much of nineteenth-century and early twentieth-century race relations, especially in the American South, since then the most common form in the United States has been aversive racism. The shift is one from racism at the level of discursive to practical consciousness. Explicit theories of white or Anglo-Saxon supremacy, although present in Bedford and Mount Kisco as elsewhere in the United States, are marginalized today. Today few people would admit to being racist, however, Young claimed that aversive racism is widespread.
11. David Sibley used the term *moral panic* to describe a situation in which a group defined as different destabilizes the social or moral order. Such panics tend to erupt when spatial and social boundaries are threatened and are often heightened by alarmist media coverage. Sibley said, "Moral panics articulate beliefs about belonging and not belonging, about the sanctity of territory and the fear of transgression. Such panics bring boundaries into focus by accentuating the differences between the anxious guardians of mainstream values and excluded others" (1995, 43). Davis (2000, 109) described the panic associated with the visibility of "street corner labor markets in edge cities and exurbs across the country" as "a nativist hysteria that frequently uses an occult pitch." On moral panics, also see Cohen (1972) and Cresswell (1996).
12. The most extreme of the many nativist reactions recorded in the 1999 public opinion survey for the new Mount Kisco Development Plan accuse immigrants of undermining the integrity of national space and of threatening the United States through various means, including terrorism. One respondent went so far as to say that since the bombing of the World Trade Center in 1993, he feels he cannot trust any immigrants, any one of whom he sees as a potential terrorist. How and why he should distinguish among them is of no consequence to him.
13. For some, the "enemy within" is no longer the communists or African Americans but the growing Latino population that transgresses the comfortable integrity and identity of the receiving community and presages the decline of Anglo or white American hegemony.
14. Baxandall and Ewen (2000, 239) said of the recently immigrated Central and South Americans, "Unlike their turn-of-the-century predecessors, these immigrants were not of one class. They were wealthy, educated, middle class, working class, uneducated, and poor." The immigrants who are most visible and cause concern in the village are the poor, male day laborers. Those with permanent work and middle-class families are, therefore, not the topic of our concern here.
15. Mitchell (1995, 1996, 1997, 1998a, 1998b) in a series of articles on law and public space discussed how the poor are increasingly being marginalized in their use of public spaces.

16. As Valle and Torres (2000, 13) posited, "Class is far more important than the specious concept of 'race' in determining the life chances of Latinos in Los Angeles."
17. As Young (1990, 146) put it,

> There exists a dissonance between group-blind egalitarian truisms of discursive consciousness and the group-focused routines of practical consciousness. This dissonance creates a sort of "border crisis" ripe for the appearance of the abject. Today the Other is not so different from me as to be an object. … But at the level of practical consciousness they are affectively marked as different. In this situation, those in the despised groups threaten to cross over the border of the subject's identity because discursive consciousness will not name them as completely different. … The face-to-face presence of these others, who do not act as though they have their own "place," a status to which they are confined, thus threatens aspects of my basic security system, my basic sense of identity, and I must turn away with disgust and revulsion.

18. As Lipsitz (1995) pointed out, "As the unmarked category against which difference is constructed, whiteness never has to speak its name, never has to acknowledge its role as an organizing principle in social and cultural relations." On the reproduction of structural privileges for whites, see Lipsitz (1998). On the spaces of whiteness, see Dwyer and Jones (2000), Kobayashi and Peake (2000), and Bonnett (1997, 2000).
19. See Duncan and Duncan (2004, chap. 8) on the battles in and out of court between landlords, town officials, and representatives of the renters.
20. On the history of the use of public space in Central America, see Low (2000).
21. The survey consisted of sixteen questions about problems facing the Village and its future growth. Some of the questions were multiple choice, with "other" as a choice and a blank to fill in if desired for some of the questions. All of the answers explicitly referring to illegal immigrants and loitering were contributed by the respondents rather than suggested by the Village as choices. However, the questionnaire did ask the following: "Do you think there is a good Quality of Life in Mount Kisco? Yes_ No_ Please identify what factors contribute to or detract from the Quality of Life" (capitalization in the original). This seems to be a somewhat leading question given the frequent use in the local media of the term "quality of life" as a code word alluding to the immigrant population. It was in fact in answer to this question that many people mentioned illegal immigrants, loitering, and illegal apartments.
    It is of course not possible to identify the ethnicity of the respondents, but it is quite likely that there were very few Latinos among those who answered the questions. We were surprised to find that the Village was not interested enough in the opinions of Latinos to have the questionnaire translated into Spanish, thereby reducing the number of Latinos who could potentially answer. Although many of the respondents were concerned about overcrowded housing, the way the problems were expressed (illegal apartments,

illegal housing, illegal tenants, deterioration of neighborhoods, nonenforcement of housing codes) suggests that these were not people who lived in overcrowded housing.

22. We should point out that included among those perceived as "rough types" are a few black teenagers as well as Latinos. Some Latinos have told us they are afraid of some of the blacks who hang out in front of a food store across from the station. Clearly there is a class as well as a race issue here.
23. See Duncan and Duncan (2004, chap. 8).

References

Agnew, J. 2005. Space-place. In *Spaces of geographical thought,* ed. P. Cloke and R. Johnson, 80–96. London: Sage.

Atkinson, D., and E. Laurier. 1998. A sanitised city? Social exclusion at Bristol's 1996 International Festival of the Sea. *Geoforum* 29:199–206.

Balibar, E. 1991. Racism and nationalism. In *Race, nation, class: Ambiguous identities,* ed. E. Balibar and I Wallerstein, 37–67. London: Verso.

Baxandall, R., and E. Ewen. 2000. *Picture windows: How the suburbs happened.* New York: Basic Books.

Belluck, P. 2005. Town uses trespass law to fight illegal immigrants. *New York Times,* July 13.

Berger, J. 1993. *Bienvenidos a los suburbios*: Increasingly New York's outskirts take on a Latin accent. *New York Times,* July 29.

Bonnett, A. 1997. Geography, "race," and whiteness: Invisible traditions and current challenges. *Area* 29:193–99.

———. 2000. *White identities: Historical and international perspectives.* Harlow, England: Prentice Hall.

Bourdieu, P. 1984. *Distinction: A social critique of the judgement of taste.* Trans. R. Nice. Cambridge, MA: Harvard University Press.

Carvajal, D. 1993. New York suburbs take on a Latin accent. *New York Times,* July 29.

Clark, F.P. 2000. *Village/Town of Mount Kisco comprehensive development plan.* New York: F.P. Clark and Associates.

Cohen, S. 1972. *Folk devils and moral panics.* London: MacGibbon and Kee.

Cooper, M. 1999. Laborers wanted but not living next door. *New York Times,* November 28.

Cosgrove, D. 1993. *The Palladian landscape: Geographical change and its representations in sixteenth century Italy.* Leicester, England: Leicester University Press.

Cresswell, T. 1996. *In place/out of place: Geography, ideology and transgression.* Minneapolis: University of Minnesota.

Daniels, S. 1993. *Fields of vision: Landscape imagery and national identity in England and the United States.* Princeton: Princeton University Press.

Davis, M. 2000. *Magical urbanism: Latinos reinvent the U.S. city.* New York: Verso.

Devine, P. 1989. Stereotypes and prejudice: Their automatic and controlled components. *Personality and Social Psychology* 56:5–18.

Duncan, J.S. 1973. Landscape taste as a symbol of group identity: A Westchester County village. *Geographical Review* 63:334–55.

———. 1999. Placelessness. In *The dictionary of human geography*, 4th ed., ed. R.J. Johnston, D. Gregory, G. Pratt, D.M. Smith, and M. Watts. Oxford: Blackwell.

Duncan, J., and N. Duncan. 2004. *Landscapes of privilege: The politics of the aesthetic in an American Suburb.* London: Routledge.

Dwyer, O.J., and J.P. Jones III. 2000. White socio-spatial epistemology. *Social and Cultural Geography* 1:209–22.

Dyer, R. 1997. *White.* London: Routledge.

Firey, W. 1945. Sentiment and symbolism as ecological variables. *American Sociological Review* 10:140–48.

Frankenberg, R. 1993. *White women, race matters: The social construction of whiteness.* London: Routledge.

Graham, B. 1994. No place of mind: Contested Protestant representations of Ulster. *Ecumene* 1:257–82.

Gross, J. 2000a. For Latino laborers, dual lives: Welcomed at work, but shunned at home in suburbs. *New York Times,* January 5.

———. 2000b. Warm smiles in strange land of the S.U.V.'s: Immigrants find friendship in suburb known for hostility. *New York Times,* July 10.

Harvey, D. 1989. *The condition of postmodernity: An inquiry into the conditions of cultural change.* Oxford: Blackwell.

———. 1996. *Justice, nature and the geography of difference.* Oxford: Blackwell.

Higley, S.R. 1995. *Privilege, power and place: The geography of the American upper class.* Lanham, MD: Rowman and Littlefield.

Hugill, P.J. 1986. English landscape tastes in the United States. *Geographical Review* 76:408–23.

———. 1989. Home and class among an American landed elite. In *The power of place: Bringing together geographical and sociological imaginations,* ed. J.A. Agnew and J.S. Duncan, 66–80. Boston: Unwin-Hyman.

Kobayashi, A., and L. Peake. 2000. Racism out of place: Thoughts on whiteness and antiracist geography in the new millennium. *Annals, Association of American Geographers* 90:392–403.

Kovel, J. 1984. *White racism: A psychohistory.* New York: Columbia University Press.

Kristeva, J. 1982. *Powers of horror: An essay on abjection.* Trans. Leon Roudiez. New York: Columbia University Press.

Kutz, C. 2000. *Complicity.* Cambridge: Cambridge University Press.

Lawrence, C. 1987. The id, the ego, and equal protection: Reckoning with unconscious racism. *Stanford Law Review* 39:317–88.

Ley, D. 1993. Past elites, present gentry: Neighborhoods of privilege in Canadian cities. In *The changing social geography of Canadian cities,* ed. L. Bourne and D. Ley, 214–33. Montreal: McGill-Queens University Press.

———. 1995. Between Europe and Asia: The case of the missing sequoias. *Ecumene* 2:185–210.

Lipsitz, G. 1995. The possessive investment in whiteness: Racialized social democracy and the "white" problem in American studies. *American Quarterly* 47:369–87.

———. 1998. *The possessive investment in whiteness: How white people profit from identity politics.* Philadelphia: Temple University Press.

Low, S. 2000. *On the plaza: The politics of public space and culture.* Austin: University of Texas Press.

Lowenthal, D. 1991. British identity and the English landscape. *Rural History* 2:205–30.

Lowenthal, D., and H. Prince. 1965. English landscape tastes. *Geographical Review* 55:186–222.

Mains, S. 2000. An anatomy of race and immigration politics in California. *Social and Cultural Geography* 1:143–55.

Massey, D. 1991. A global sense of place. *Marxism Today,* June, 24–29.

———. 1993. Power/geometry and a progressive sense of place. In *Mapping the futures,* ed. J. Bird et al., 59-69. London: Routledge.

———. 2005. *For space.* London: Sage.

Matless, D. 1998. *Landscape and Englishness.* London: Reaktion Books.

McIntosh, P. 1988. White privilege and male privilege: A personal account of coming to see correspondences through work in women's studies. In *Wellesley College Center for Research on Women Working Paper Series* 189. Reprinted in Andersen, M., and P. Hill, eds. 1992. *Race, class, and gender: An anthology,* 70–81. Belmont, CA: Wadsworth.

Mendus, S. 2000. *The politics of toleration in modern life.* Durham: Duke University Press.

Mitchell, D. 1995. The end of public space? People's Park, definitions of the public, and democracy. *Annals, Association of American Geographers* 85:108–33.

———. 1996. Political violence, order and the legal construction of public space: Power and the public forum doctrine. *Urban Geography* 17:152–78.

———. 1997. The annihilation of space by law: The roots and implications of anti-homelessness laws in the United States. *Antipode* 29:303–35.

———. 1998a. Anti-homeless laws and public space: I; Begging and the First Amendment. *Urban Geography* 19:6–11.

———. 1998b. Anti-homeless laws and public space: II; Further constitutional issues. *Urban Geography* 19:98–104.

Nevins, J. 2002. *Operation Gatekeeper: The rise of the "illegal alien" and the making of the U.S.-Mexico boundary.* New York: Routledge.

Ortiz, P. 2000. Day laborers rattle Valley communities: Are sidewalks big enough for all? *Arizona Republic,* December 6.

Philo, C. 1998. A lyffe in pyttes and caves: Exclusionary geographies of the West Country tinners. *Geoforum* 29:159–72.

Pile, S. 1994. Masculism, the use of dualistic epistemologies, and third spaces. *Antipode* 26:255–77.

Pratt, G. 1981. The house as expression of social worlds. In *Housing and identity: Cross-cultural perspectives,* ed. J. Duncan, 135–80. London: Croom Helm.

Price, P. 2000. Inscribing the border: Schizophrenia and the aesthetics of Aztlan. *Social and Cultural Geography* 1:101–16.

Pulido, L. 2000. Rethinking environmental racism: White privilege and urban development in Southern California. *Annals, Association of American Geographers* 90:12–40.

Purdy, M. 2001. Laborers wanted, sometimes. *New York Times,* April 15.

Rose, G. 1993. *Feminism and geography: The limits of geographical knowledge.* Cambridge: Polity.
———. 1995. Place and identity: A sense of place. In *A place in the world? Place, cultures, and globalization*, ed. D. Massey, and P. Jess, 87–127. Oxford: Open University Press.
Sibley, D. 1995. *Geographies of exclusion: Society and difference in the West.* London: Routledge.
———. 1998. The problematic nature of exclusion. *Geoforum* 29:119–22.
Soja, E. 1989. *Postmodern geographies: The reassertion of space in critical social theory.* London: Verso.
Toma, R. 2001. Farmingdale should learn a lesson. *Newsday,* April 8, A25.
Town of Bedford. 1992. Zoning. In *Code of the Town of Bedford, Westchester County, New York,* 12501–633. Rochester, NY: General Code.
———. 1997. *Conservation board questionnaire.* Rochester, NY, Town Hall.
———. 1998. *Conservation board questionnaire: Summary and findings.* Rochester, NY, Town Hall.
———. 1999. *Master plan questionnaire.* Rochester, NY, Town Hall.
———. 2000. *Master plan questionnaire analysis of findings.* Rochester, NY, Town Hall.
Valle, V., and R. Torres. 2000. *Latino metropolis.* Minneapolis: University of Minnesota Press.
Wyckoff, W. 1990. Landscapes of private power and wealth. In *The making of the American landscape,* ed. M.P. Conzen, 335–54. London: Unwin-Hyman.
Young, I.M. 1990. *Justice and the politics of difference.* Princeton: Princeton University Press.
Zavella, P. 2000. Latinos in the USA: Changing socio-economic patterns. *Social and Cultural Geography* 1:155–67.

CHAPTER 9

The Cultural Landscape of a Latino Community

JAMES ROJAS

As greater numbers of Latino immigrants and native-born Mexican American citizens settle into East Los Angeles, they bring with them a different use of urban space to an already existing built environment. Homes, *ciudades, pueblos,* and *ranchos* in Latin America are structured differently both physically and socially than typical American suburbs. Latinos are retrofitting the built environment of East Los Angeles to fit their cultural needs. The suburban landscape of this community has changed greatly with the arrival of Latinos. This chapter explores the burgeoning East Los Angeles community and describes and speculates on the meaning of the new spatial arrangements. Both new behavior patterns and their concomitant new visual urban order are previously unknown to builders, users, and observers of the cultural landscape in Los Angeles.[1]

Landscapes of Public Life

Latinos bring a rich practice of public life to Los Angeles that can be seen by the way they retrofit the urban street design. Street vendors carrying their wares, pushing carts, or setting up temporary tables and tarps; vivid colors, murals, and business signs; clusters of people socializing on street corners and over front yard fences; and the furniture and props that

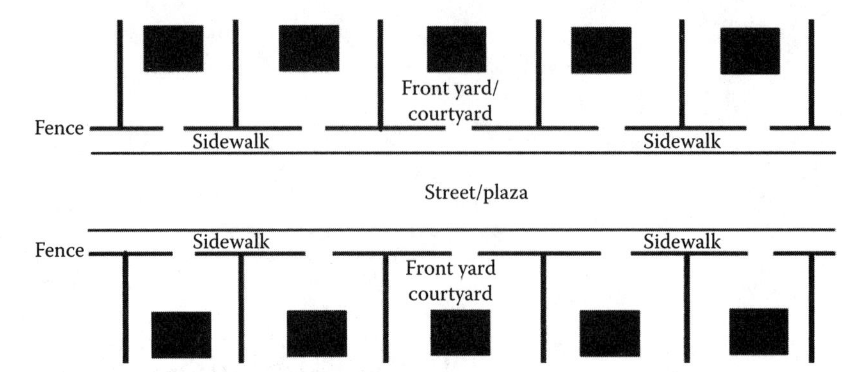

Figure 9.1 Setting the stage: the residential street pattern provides a framework for Latino practices of public life in East Los Angeles.

make these front yards into personal statements all contribute to the vivid, unique landscape of the city (see Figure 9.1).

Very few signs or landmarks indicate the precise location of East Los Angeles. However, you know when you have arrived there because of the large number of people on the streets engaging in all types of activities. What may look like random groups of people are actually well-ordered interactions in which everybody has a role. Children play, teenagers hang out, and older people watch. These people in these roles enhance the street activity and provide security for their families, neighbors, and friends. People on the streets in East Los Angeles exercise implicit social control. Everybody knows each other in their neighborhood. If you do not belong, they will challenge you with words or a stare.

Street Vendors

The streets of East Los Angeles resemble the streets of Latin America in large part because of the presence of street vending. The streets provide for economic survival by giving Latinos a place to sell items and their labor. Economic survival emphasizes the importance of the street and changes the urban landscape by adding more activity to the scene. In Mexican cities, outdoor selling is very popular. Latinos have transferred this practice to the United States. However, much of the American urban landscape is not designed for outdoor selling. In East Los Angeles, vendors have had to be very innovative in adopting ill-designed urban space to their needs. For example, vendors selling on the exits and entrances of freeways transform a space that was not designed for pedestrians. Vendors also occupy intersections and temporarily transform vacant lots, front yards, sidewalks, and curbs into ephemeral markets.

East Los Angeles vendors vary in their mobility and amount of investment. One type is highly mobile and has little overhead. Day laborers station themselves near hardware stores throughout Los Angeles, competing for menial jobs that are in short supply. If drivers show any sign of interest and slow down even slightly, they will be swarmed by work-hungry men. Many *mariachi* bands hang out at the intersection of First and Boyle, now referred to as *Mariachi* Plaza, where they wait patiently for people to drive up and hire them for parties. Other *mariachis* walk the streets of East Los Angeles going from bars to restaurants serenading patrons. Other men walk the streets carrying fake flowers or cassette tapes or other light items to sell. Still other men and women station themselves at freeway off-ramps or median strips at major street intersections, sometimes carrying their wares.

Others selling similar products create tents as protection from the sun and rain by ingeniously attaching sheets of plastic or fabric to buildings, poles, and fences. In the past few years there has been an increase of small barbecue food stands known as *taquerías,* mainly on commercial streets in East Los Angeles. On the weekends many people also sell odds and ends in their front yards. Fences are an important part of this composition because they can hold up items and delineate the selling space. This adds significant activity to many residential streets.

Still others circulate through the neighborhood with carts, cars, and trucks. Pushcart vendors roam residential and, to a lesser extent, commercial streets of East Los Angeles, primarily selling food. Many walk the same beat day in and day out. Pushcart vendors sometimes look out of place on the wide suburban streets of Los Angeles because the streets are designed for cars. Car and truck vendors conduct business from the side of their vehicles or park along streets or in vacant lots and set up tents. Some move several times a day. Large food trucks, traditionally part of the American scene, are now in East Los Angeles, called "roach coaches" because they are known to be somewhat dirty. They provide lunches at many construction sites, dinners in East Los Angeles, or late night snacks when the bars close down at 2:00 a.m.

Street vendors have been a part of the enacted environment in Latino East Los Angeles since the 1920s. However, with the rising Latino population in the past few years, there has been a drastic increase in street vending. In some communities around East Los Angeles, this type of selling is prohibited. Furthermore many residents complain about the trash these vendors generate, while permanent businesses complain about unfair competition.

Props

No space in East Los Angeles is left unused or unmarked. Moveable props help create the enacted landscape of East Los Angeles. A table and chair brought outside define territory and make people comfortable in the open urban space. A parked car can also become the center of a day's activity for men working on the street or in the front yard. A sofa under a tree or on a porch allows residents to while away the afternoon, while a barbecue grill can generate some revenue and become a focal point for neighborhood gossip. From car stereos to *mariachis,* music in East Los Angeles also serves as a prop that temporarily controls and defines space.

Many shopkeepers in East Los Angeles have removed the glass front wall of their shops, which open to the street during business hours. Wares are placed outside on the sidewalk, creating a flow between the indoor and outdoor spaces. Shop wares, racks of clothes, pushcarts, and parked cars take over the sidewalks and streets, creating a tactile environment in front of these stores.

Paint also serves as a prop, which allows Latinos to quickly redefine the cultural landscape. Very few spaces and walls are left untouched in East Los Angeles. From graffiti and store signs to murals, walls become a medium for Latino cultural expression.

In Mexico, murals and graphics are important communication tools that date back to the days when the Spanish arrived in Latin America and had to communicate with the indigenous people. Pictures and words were used to give directions in Mexico City. For example, to this day a pig or cow head indicates a butcher shop, while a cornucopia indicates a vegetable and fruit stand. The flamboyant words and graphics covering many buildings from top to bottom add a kinetic visual element to the urban environment.

Murals used for business advertisement can also be political or religious. Murals liven up a space. Murals painted on the large sidewalls of businesses facing residential streets not only advertise products and services and disguise graffiti but also create a transition from commercial to residential uses. The use of props in both residential and commercial areas in East Los Angeles creates a connection between the two zones. Props help to humanize the scale of interaction in the commercial landscape by bringing that interaction to a pedestrian level, which contradicts the automobile scale of Los Angeles. Driving through the streets of the Latino *barrios*, all ones sees is clutter. Walking, however, one experiences a rich tactile landscape that enhances the enacted environment.

Fences: a Social Catalyst

Waist-high fences are ubiquitous throughout the residential landscape of East Los Angeles. They outline most front yards and define the streetscape of Latino barrios. Most are visually permeable chain link or elaborate wrought iron. In other, non-Latino neighborhoods, people build fences for security and privacy, as they do in East Los Angeles. But in middle-class suburban neighborhoods, people rarely congregate in the front yard. This visible expanse of the front yard acts as a psychological barrier separating the private space of the home from the public space of the street. However, in East Los Angeles fences serve other purposes. From a place to hang wet laundry, to focus neighborly talk, and to display items for sale, fences are a useful threshold between the household and the public domain.

In Latino neighborhoods, fences define boundaries between public space and private space. They create easily defendable spaces and assert ownership. But, they also break down social barriers by creating a place where people can congregate. Fences serve as social catalysts. The threshold of the home is a pivotal part of the household. It marks the boundary between an open, public realm and a closed and inaccessible private one: the boundary between outside and family.

Visitors rarely cross the threshold of a home unless invited to do so. Inviting or not inviting someone to enter the home is a clear signal of the occupant's desire for more or less contact. When a resident and visitor meet at a door threshold, there is great pressure to define the social relation: will the visitor be invited in or not? But in East Los Angeles the fence around a front yard moves the threshold from the front door to the front gate. This removes the pressure of social definition, making it acceptable to entertain people informally over the front yard fences.

Taken together, the enclosed front yards of East Los Angeles create a different urban landscape. These fences transform the street into an unconventional and unofficial plaza through their social use. Residents and pedestrians participate in social dialogue on the street through the comfort and security of that visible threshold between their enclosed front yards and the public street.

***La Yarda*: a Personal Expression**

The enclosed front yard is the cultural landscape element that especially translates the Mexican use of space to East Los Angeles, while also expressing personal and family identity. Here the residents put their faces on the street. The fences clearly delineate an area residents can personalize without interfering with their neighbors. Enclosed front yards function as a work space, a party area, or just a place to hang out. They vary from elaborate gardens reminiscent of Mexican courtyard houses to unkempt yards. Front yards reflect the Mexican cultural values applied to American suburban form.

Support of community identity is measured by how well the household upholds the neighborhood standards through the upkeep of the front lawn. A balance is struck between the collective and the individual identity in the front yard.

The homes in East Los Angeles were built by non-Latinos but have evolved into a vernacular form as residents have made changes to suit their needs. Every change, no matter how small, has meaning and purpose. Bringing the sofa out to the front porch, stuccoing over the clapboard, painting the house vivid colors, or placing a statue of the Virgin in the front yard all reflect the struggles, triumphs, and everyday habits of working-class Latinos. The front yards in East Los Angeles are not anonymous spaces but personal vignettes of owners' lives.

Unlike middle-class suburban houses that seem to be insulated from the street, East Los Angeles houses extend graciously to the street. Each house communicates with the street and other houses through the use of fences and props, and so houses and street are drawn together. A hybrid of two architectural vocabularies, Latino homes and barrios create a new language that uses building elements from both Mexico and the United States. The house may still be freestanding as in the typical American suburb, but the new fences create a semiprivate space for the family drawn from the interior courtyards of Mexican houses (see Figure 9.2).

Upon entering one of these enclosed front yards, the resident's private world unfolds. What appears cluttered from outside the fences becomes as organized as the objects in a room: the enclosed front yard acts like a

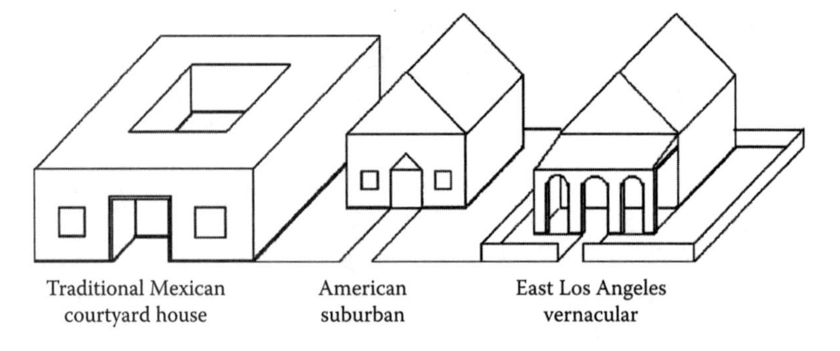

Figure 9.2 Evolution of an East Los Angeles vernacular housing style: public–private interaction is mediated through the enclosed front yard.

room belonging to the house. In this space, all the sights and sounds from the street are tailored to the needs of the owner. Mexicans bring the party, work space, and conversation to the front yard.

The front porch becomes one of the focal points of the house because of all the activity in the front yard. In most American homes today, the use and importance of the front porch has declined for various reasons. However, in East Los Angeles front porches have gained a new importance as residents enlarge and expand them for heavy use. Residents sit on porches to escape summer heat or just to be outside with family, friends, and neighbors.

The Politics of Urban Space

My research on the cultural landscape and use of space by Latinos in East Los Angeles has led to the development of the Latino Urban Forum (LUF). The LUF examines the built form as an innovative approach to community organizing and design. This approach has been very successful because land use is a fundamental issue in the dense neighborhoods inhabited by many Latinos. The LUF helps residents envision improvements to their community by providing technical skills, making policy changes, and helping residents articulate their concerns. The LUF is the only organization in Los Angeles that exclusively examines the issues of urbanism and the planning process and advocates for an enhanced built environment as it relates to underserved Latinos. The LUF is concerned with issues identified by the community and the processes necessary to resolve related problems. In doing so, the LUF enlists the assistance of influential Latinos wielding the ability help those who are lacking in resources.

Latinos, Parks, and Open Space

The LUF has developed an expertise for examining and creating open space projects in East Los Angeles. Plazas in Latin America are used for all types of activities, from vending to socializing to recreating during a twenty-four-hour day. These same activities occur in streets, vacant lots, and other open areas of East Los Angeles, including the enclosed front yards, which contribute to redefining "public life" on the streets. In East Los Angeles residents come together daily for many activities and celebrations, especially during the Christmas season with the *posadas* and the creation of Nacimientos (highly elaborate nativity scenes).

The few parks that already exist are heavily used for family celebrations, *quinceanera* photographs, and active recreation. Many of these parks were designed at the turn of the century for sedentary and pensive activities and lack facilities for these new activities. Latinos create their own recreation space from the existing streets and sidewalks. Children on the streets play many games including tag, football, soccer, and basketball. The LUF's interest in the changing urban landscape and its active role in redefining urban space has led to at least two ongoing public projects in East Los Angeles: the Evergreen Cemetery jogging path and the South Central Farm, an urban garden project.

Evergreen Cemetery Jogging Path

The Evergreen Cemetery jogging path is an innovative, urban, recreational space project conceived by the LUF in collaboration with neighborhood residents and local government. A 1.5-mile-long city sidewalk in the Boyle Heights community of Los Angeles has been converted into a rubberized jogging path for community use. This path serves as a multifunctional open recreational space, a public right of way, and an urban green space. The Evergreen jogging path serves as a public space of community pride, promoting social interaction, creating neighborhood goodwill, and encouraging good health. Today more than one thousand joggers and walkers use this facility each day.

South Central Farm

The LUF has been providing organization assistance and media consultation to the urban farmers of the South Central Farm to save the fourteen-acre community garden at 41st and Alameda from becoming a warehouse. For the past twelve years, through their hard labor, the gardeners have created a magical community asset by creating a lush green space in their concrete urban environment. The gardeners have built fences around their plots, tilled and fertilized the land by introducing organic matter, and planted trees, which attract birds and butterflies. Narrow pathways separate the 318 garden plots in a maze of foliage reminiscent of the high hedge gardens of Europe. Vines grow on the chain-link fences and form canopies over the pathways. The same vines grow into the gardens and provide shade for the gardeners. The dense foliage creates a junglelike damped-air quality. The gardeners grow many crops not found in the American marketplace, making the garden a horticultural exploration.

This garden is a self-sufficient economic asset for the community. About 350 low-income gardeners grow edible crops that can easily reduce a family grocery bill by one-third. The weather in Los Angeles allows for up to four productive harvests through aggressive farming techniques that involve composting and eliminating stubborn weeds, pesky pathogens, and fungi.

The community at large benefits because the supplemental crops are sold to local residents at prices lower than those at local markets. On weekends visitors walk the narrow pathways looking for hard-to-find produce. Vendors come and sell ice cream and drinks at the garden on weekends. The gardeners do not get public funding.

The garden is a safe haven, a drug-free, gang-free, and graffiti-free area. Residents from this community and beyond visit this urban oasis to find peace of mind and to relax with family and friends. The garden creates strong social networks and reinforces family values. For the past ten years, many gardeners and their children have grown up there and know everybody on a first-name basis. Children run freely and safely though the narrow pathways, while their parents and grandparents tend their plots. Children also learn from their parents and grandparents the beauty and importance of gardening.

The garden gets as much use as many city parks and has even established its own traditions, like the celebration for Garden Day L.A., when one hundred food and plant vendors, musicians, and folklorica dancers created a one-day fair at the garden. The South Central Farm is unique, and it is necessary in this community. Many Latinos come from rural backgrounds in Latin America, and at the South Central Farm they can find a community open space that meets their needs.

Conclusion

The growing Latino population has created a new cultural landscape out of the existing suburban form of East Los Angeles. Street vendors carrying their wares, pushing carts, or setting up temporary tables and tarps; vivid colors, murals, and business signs; clusters of people socializing on street corners and over front yard fences; and the furniture and props that make these front yards into personal statements all contribute to the vivid, unique landscape of the barrio. Residents have focused their inventive imaginations on the transformation of the spaces in front of and between structures. The cultural behavior patterns of Latinos render the city's current land-use practices out of time and out of place for protecting and promoting the well-being of the community. Changes in the East Los Angeles cultural landscape have been wrought through individual adaptation and creativity as well as through the concerted efforts of the community at large, including the actions of the LUF.

Note

1. This chapter is drawn from a larger study: James Rojas, "The Enacted Environment: The Creation of Place by Mexicans and Mexican Americans in East Los Angeles" (master's thesis, MIT, 1991).

CHAPTER 10

The Witting Autobiography of Richmond, Virginia: Arthur Ashe, the Civil War, and Monument Avenue's Racialized Landscape

JONATHAN LEIB

> [Richmond chief of police Judy] Hammer was extremely sensitive to racial issues and had studied the Richmond metropolitan areas thoroughly. She knew it wasn't so long ago that blacks couldn't join various clubs or live in certain neighborhoods. They couldn't use golf courses or tennis courts or public pools. Change had been slow and in many ways deceptive.
>
> Membership and neighborhood associations began to accept blacks, and in some cases women, but making it off the waiting list or feeling comfortable was another matter. When the future first black Governor of Virginia tried to move into an exclusive neighborhood, he was turned down. When a statue of Arthur Ashe was erected on Monument Avenue, it almost caused another war.[1]

Best-selling author Patricia Cornwell's 1999 mystery novel *Southern Cross* revolves around issues of race relations in Richmond, Virginia. In the book, Cornwell focuses in part on issues of race and Richmond's Civil War legacy. One of the book's subplots revolves around the vandalism of a statue of Confederate president Jefferson Davis in the city's historic Hollywood Cemetery, where Davis is repainted to resemble an African American University of Richmond basketball player. The book's title, *Southern Cross*, is a reference to the best-known symbol of the Confederacy, its battle emblem (shown on the book's cover in case readers missed the reference), itself the subject of much racial controversy over the past decade.[2]

As the former capital of the Confederate States of America, Richmond's landscape has the densest concentration of memorials to the Confederacy of any large Southern city, ranging from a sixty-foot-tall monument to Robert E. Lee to a downtown street named after his horse Traveller. However, of all the tributes to the Confederacy, the grandest is the city's Monument Avenue, along which statues of Confederate heroes (among which are three of what Charles Reagan Wilson labeled the "Lost Cause Trinity" of Lee, Davis, and Stonewall Jackson)[3] were built in the late nineteenth century and early twentieth century. A century later, a National Park Service publication proclaimed Monument Avenue as "the South's grandest commemorative precinct dedicated to the heroes of the Lost Cause."[4]

However, as Patricia Cornwell pointed out, in 1995 "another war" was almost caused by a vociferous debate in Richmond over erecting a sixth statue on Monument Avenue, the first in more than six decades. This time, the hero honored was to be the late Arthur Ashe, a Richmond native, a tennis legend, a noted human rights activist, and, important to this debate, an African American. The debate over whether a statue of Ashe should join those of (white) Confederate heroes on Monument Avenue divided the city and raised numerous questions concerning race in Richmond and within its symbolic landscape.[5]

In this chapter we examine the debate over placing a statue of Arthur Ashe on Richmond's Monument Avenue. First, we briefly examine the history of Monument Avenue, interpret its landscape over time, and discuss how both white and black Richmonders have viewed and shaped its symbolism and how actions vis-à-vis Monument Avenue have both reflected and challenged the city's race relations. With this backdrop, we then examine the 1995 debate over placing a statue of Arthur Ashe on Monument Avenue. The debate over desegregating Monument Avenue raised questions about the meaning of Ashe, the Confederacy, and Monument Avenue and highlighted and challenged Richmond's white racialized landscape.

A Century of Making and Remaking a White Racialized Landscape: Monument Avenue, 1890–1992

Constructed west of downtown Richmond beginning in the late nineteenth century, Monument Avenue was built as a grand residential avenue, similar

Figure 10.1 Mid-twentieth-century postcard showing a view of Monument Avenue, facing west (from the collection of Jonathan Leib).

to other grand avenues built at the time in cities such as Chicago, Cleveland, and Washington, D.C.[6] A century later, Monument is still a grand avenue, and for much of the past century it has been the city's most famous and fashionable street. As Busbee noted, Monument Avenue is today "one of the South's grandest streets, an impressive stretch of red brick Colonial Revival homes divided by an oak-lined median" (see Figure 10.1).[7] In the early twentieth century, as Richard Guy Wilson suggested, Monument Avenue "was the preferred address for the local wealthy and those who aspired to that status."[8]

In terms of the avenue's symbolism, if, as Peirce Lewis famously wrote, "the human landscape is our unwitting autobiography, reflecting our tastes, our values, our aspirations, and even our fears in tangible visible form," Monument Avenue quite consciously reflects a *witting* autobiography of the white South in the late nineteenth century and early and mid-twentieth century. Monument Avenue's witting autobiography is that of a prime example of a white racialized landscape.[9] As Schein suggested, a "cultural landscape can be racialized, and a racialized landscape serves to either naturalize, or make normal, or provide the means to challenge racial formations and racist practices."[10] The whiteness of Monument Avenue's landscape is unmistakable in the impressive monuments that were built that gave the avenue its name. From 1890 to 1929, statues of five Confederates were built on Monument Avenue—Robert E. Lee in 1890, Jefferson Davis and J.E.B. Stuart in 1907, Stonewall Jackson in 1919, and Mathew Fontaine Maury (best known for his work in oceanography but with Confederate connections) in 1929—as many large homes (owned by whites) were also being built. The monuments and the avenue were both constructed at the height of the Lost Cause era, when Southern whites were (re)writing the history of the Confederate cause in the Civil War as a noble struggle against Northern aggression rather than as a losing effort to preserve the institution of slavery, and at the same time as Jim Crow laws and practices were going to into effect in Richmond and elsewhere in the South that spatialized white domination over African Americans.[11] The statues to Confederate heroes along Monument Avenue were one focal point of this effort to script white Southern identity on the landscape (this era, the late nineteenth century and early twentieth century, marked the period when most Confederate monuments were erected in the South). Richard Guy Wilson highlighted the importance of the statues, noting that Monument Avenue "may be a place of residences and churches, a street of movement and communication, but ultimately Monument Avenue is the site of memorials to the Confederacy. And it is back to these statues one must come, for their message cannot be ignored."[12]

The whiteness of the avenue's landscape has been evident in the building of the monuments and various proposals for the avenue since its beginning. The first statue built was a sixty-foot monument to Confederate general Robert E. Lee, "the white South's favorite icon" (see Figure 10.2).[13] As Leib noted, "Lee became the most important symbol of the late-nineteenth-century's Lost Cause Movement," with Thomas Connelly suggesting that "Lee became a God figure for Virginia."[14] The unveiling ceremony in 1890, which DuPriest and Tice suggested was "one of the grandest spectacles in Richmond history," drew a crowd estimated at more than one hundred thousand.[15] The worship of Lee in Richmond was evident in that in moving the statue into its final Monument Avenue location, most of Richmond's

> white citizenry [was] involved in transporting the statue through the streets. Infants and toddlers were taken from the nursery to touch the ropes that pulled the statue; one of four ropes was especially for young ladies, and pieces of the rope were kept as souvenirs and passed down in families.[16]

The Monument Avenue unveiling ceremonies for the Stuart and Davis monuments in 1907 and the Jackson monument in 1919 provided more opportunities to revere and celebrate the Confederacy. Indeed, throughout this period, Richmond in general and Monument Avenue in particular were the focal points for numerous Confederate parades and celebrations. As Charles Reagan Wilson suggested about the development of the Myth

Figure 10.2 Robert E. Lee monument, Monument Avenue and Allen Avenue.

of the Lost Cause, "Richmond was the Mecca of the Lost Cause, and Monument [Avenue] … was the sacred road to it."[17]

However, from its beginning, the whiteness of the Monument Avenue landscape did not go unchallenged by the city's African Americans. Prior to being disenfranchised in the early twentieth century, Richmond's city council had several black members who refused to vote (though they were outvoted) for the appropriation of city funds for the 1887 ceremony laying the cornerstone for the Lee Monument or its 1890 dedication.[18] City councilman and black community leader John Mitchell Jr. argued that taxpayer money should not be used to honor Lee, noting that the Lee statue dedication ceremony, rather than being a celebration of a glorious Lost Cause, "handed down a legacy of treason and blood" to future Richmonders.[19] Indeed, Mitchell reported that at the 1890 monument dedication ceremony, one African American noted that the imposing statue signified, in terms of Southern social relations, that, literally, "the Southern white folks is [*sic*] on top."[20]

Monument Avenue continued as an overt symbol of white power through the mid-twentieth century. Reflecting on his childhood in 1950s Richmond, Arthur Ashe later noted that the city's most influential white church, First Baptist, "confirmed its domination and its strict racial identity by its presence on Richmond's Monument Avenue, the avenue of Confederate heroes."[21]

Proposals for Monument Avenue in the 1960s intersected with and symbolized the struggle for and resistance to civil rights in Richmond. After World War II, many whites in Richmond began leaving the city for the suburbs, and while the neighborhood surrounding Monument Avenue was still majority white, the avenue was considered to be in decline.[22] In 1965, as the city was in the midst of a major struggle over school desegregation and parts of black neighborhoods had been or were proposed for demolition to promote "urban renewal," the Richmond City Planning Commission issued a plan to refurbish Monument Avenue, including extensive landscaping and the building of seven more statues to Confederate heroes.[23] Among the potential new statues was one proposed by Salvador Dali that would have honored Confederate nurse Sally Tompkins, which would have featured, in the words of Driggs, Wilson, and Winthrop, a sword-wielding Tompkins "as a latter-day St. George, slaying a germ-dragon atop a pedestal composed of a Petri dish balanced on a finger."[24] Although the statues, including Dali's, were never sculpted, the message sent by the Planning Commission was clear: at a time when whites were leaving the city and the city was in the midst of the civil rights struggle, the proposal to build more monuments to an earlier Lost Cause was a call to whites to remain. That same year, however, a protest was lodged against the Confederate statues on Monument Avenue when vandals painted the faces of both Lee and Davis black.[25]

In the late 1960s, as the percentage of Richmond's white population declined and the percentage of its African American population increased, there was concern among some Richmond whites that African Americans would shortly become a majority of the city's population and, if African Americans voted as a block, gain political control of the city for the first time. The concern was so great that in 1970 the city's white leadership annexed an overwhelmingly white portion of a neighboring county to dilute the percentage of the city's black population to ensure that whites remained a majority. A lawsuit was ultimately successful in overturning the annexation on the grounds that under the federal Voting Rights Act the annexation illegally diluted black voting strength.[26] Reminiscent of John Mitchell Jr.'s vow in 1890 that "the Negro put up the Lee Monument, and should the time come, will be there to take it down," one of white Richmond's concerns with African Americans potentially gaining control of the city council was that they would tear down the Confederate statues on Monument Avenue.[27] In 1968 one of Richmond's (white) state senators introduced legislation, subsequently approved and signed into law, that gave Virginia's attorney general eminent domain power to seize the Confederate monuments from the city if such action was in the "public interest" in the estimation of the attorney general.[28] The senator who introduced the bill later explained that it was necessary to preserve the monuments because at the time there were rumors that among African Americans in the city "there was a movement to remove the Confederate generals," as the monuments "were revolting to the black people of Richmond."[29] Thus, the statues on Monument Avenue were viewed as a prime symbol of white power in the city. Indeed, Barone and Ujifusa reported that in 1977, shortly before the first black-majority city council was seated, the white-majority council deeded the Lee monument to the state to prevent its possible removal.[30]

Although Monument Avenue had been considered to be in a state of decline, efforts to preserve the avenue and its historic homes began in earnest in the 1970s, with Edwards, Howard, and Prawl suggesting that for "many homeowners, living on Monument was a means of reclaiming the past, of celebrating the history of an avenue that had itself been built to commemorate an era long gone."[31] However, this is not to suggest that everyone in Richmond wishes to celebrate the Monument Avenue's history or the era it was designed to commemorate. Monument Avenue's original home owners were white, and today the avenue sits in Richmond's majority-white Fan District of this now African American–majority city. Although some see the avenue's statues as a reminder of a glorious past, black Richmond city councilman Chuck Richardson noted, in his 1991 attempt to integrate the avenue with a statue of Richmond civil rights heroes, that instead "the beautiful statues on Monument Avenue stand in substantial measure for a society which believed in subjugation of the black race."[32] Although Richardson's proposal did not gain approval, the city council the following year approved a resolution that sought to destabilize the long-regarded meaning of the avenue by stating, "Monument Avenue was not the exclusive domain of Confederate heroes."[33]

"Another War": the 1995 Debates over the Arthur Ashe Statue and the Meaning of Monument Avenue

It is within this context of race relations in Richmond and their reflection in the history of the making and remaking of Monument Avenue's meaning that the 1995 debates over where to erect a statue of Arthur Ashe must be placed. Given the history of Monument Avenue, in a city about whose white and black citizenry Barone and Ujifusa suggested in 1997 that "there remains a gulf between these two separate cultures, connected but still not unified," it is not surprising that the proposal to place a statue of Arthur Ashe on Monument Avenue exploded in a "firestorm" of vitriolic rhetoric, as proponents and opponents of the Monument Avenue site argued to control the definition of what Monument Avenue's landscape means and how placing Ashe on Monument Avenue (or keeping him off) would either alter or reinforce that meaning.[34]

Before examining the debate over placing a statue of Ashe on Monument Avenue, we must briefly discuss Ashe himself. He was born in Richmond in 1943 and was raised during the city's segregationist period before leaving Richmond and the South in 1960 to pursue his tennis and classroom education. In the 1970s Ashe became one of the most admired and successful tennis players in the country, becoming the first black male to win the U.S. Open and Wimbledon tournaments before retiring in 1979 following a heart attack. Following his retirement, Ashe successfully pursued his humanitarian and philanthropic interests. He was best known as one of the leading American activists working to end the apartheid regime in South Africa (indeed, when Nelson Mandela was released from Robben Island prison, he mentioned that Ashe was the first American he wanted to meet).[35] Ashe died in February 1993 from pneumonia that he received while suffering from AIDS (which he had acquired through a blood transfusion during heart surgery). Once his condition was revealed by the press in 1992, he gave numerous lectures on AIDS and set up a foundation aimed at fighting the disease. Though he was retired from the sport for more than a decade, in 1992 *Sports Illustrated,* recognizing his human rights efforts as well as his tennis career, named Ashe its "Sportsman of the Year."[36]

Ashe's childhood experiences living under segregation left him bitter toward Richmond. As he noted about his exodus from the city in 1960, "When I decided to leave Richmond, I left all that Richmond stood for at the time—its segregation, its conservatism, its parochial thinking, its slow progress toward equality, its lack of opportunity for talented black people."[37] Ashe's work on trying to bring about the end of apartheid was inspired by his childhood memories, noting that the "core of my opposition to apartheid was undoubtedly my memory of growing up under segregation in Richmond."[38] However, by the time of his death, Ashe had, in his own words, "made peace with the state of Virginia and the South."[39] Although he lived in New York following his retirement from tennis, he did come back to Richmond frequently, setting up the organization "Virginia Heroes" to help at-risk youth in the city. Richmond leaders recognized his efforts by naming an athletic center in his honor. When Ashe died, his body was flown to Richmond for the funeral (at the Ashe Athletic Center), while five thousand persons came to view his body as it lay in state at the Executive Mansion (the home of then-governor Douglas Wilder, a Richmond native and the country's first black governor).

Upon his death, Ashe was eulogized throughout Richmond as a hero, with numerous suggestions made as to how to honor him.[40] The year before, Richmond artist Paul DiPasquale had contacted Ashe about sculpting a statue of Ashe, and he agreed. Following Ashe's death, Virginia Heroes agreed to raise the money to make DiPasquale's statue a reality. Virginia Heroes and DiPasquale met with city officials who accepted the statue, and in December 1994 a prototype of the statue was unveiled to kick off the fund-raising efforts to complete it. The statue's design features a standing Ashe dressed in a warm-up suit, with children looking up at him. In Ashe's raised right hand are books, while in his raised left hand is a tennis racket. The books are higher than the tennis racket, which, combined with the children in the statue, emphasize Ashe's paramount belief in the importance of education (see Figure 10.3).[41] When asked before he died why he would want a statue of himself placed in the city that he had left

Figure 10.3 Arthur Ashe monument, Monument Avenue and Roseneath Road.

as a teenager because of segregation, Ashe responded, "Just because they turned their back on me, I don't have to turn my back on them."[42]

Although there was little controversy over having a statue of the universally admired Ashe on Richmond's landscape, controversy erupted over where within Richmond's landscape the statue would be placed and what that location would mean. At an early showing of the monument prototype, former governor Wilder suggested the finished statue be placed on Monument Avenue. DiPasquale later recalled, "Nobody agreed with him. … Everybody thought it would just be too much of a fight."[43] In December 1994 Wilder publicly suggested that the Ashe statue be placed on Monument Avenue, sparking the controversy over the site.[44] After considering several options, Virginia Heroes proposed a site on Monument Avenue at a June 1995 hearing of the city's Planning Commission, which was to vote on where the Ashe statue would be situated. While proposing a site on Monument Avenue, the Virginia Heroes proposed a site for the Ashe statue that lessened the potential criticism from those not wanting to see Ashe on the same street with the Confederate statues. The Confederate statues are located on the part of the avenue that was first built, the eastern one and a half miles of its five-mile length. This stretch, from Lombardy Street to Roseneath Road, the Monument Avenue historic district, is the most impressive part of the avenue; Edwards, Howard, and Prawl noted it is a "cohesive area with a consistent road width, continuous plantings, and harmonious architecture." West of Roseneath, the avenue's later addition, "becomes increasingly suburban in appearance" and is "no longer easily perceived as a majestic, in-town boulevard."[45] The Virginia Heroes' proposed site for the Ashe statue was in this newer part of the avenue, three blocks west of Roseneath and, as important, just west of where Monument Avenue crosses Interstate 195, a significant break in the avenue (see Figure 10.4). Thus, as Leib suggested, the Virginia Heroes' proposed location meant that "while Ashe would be on the same Avenue with the Confederates, he would be physically, temporally, and symbolically separated from them."[46]

However, although the Planning Commission (whose membership was made up of five white and four African American members) voted unanimously to place the Ashe statue on Monument Avenue, they changed its location from the one the Virginia Heroes had proposed. Instead of placing the Ashe statue to the west of the interstate, they moved the location three blocks to the east on the same side of Monument Avenue as the Confederate statues, to the intersection of Monument Avenue and Roseneath Road, at the edge of the historic district and inside the spot, memorialized with a monument of a cannon, of the Confederate's second line of defense during

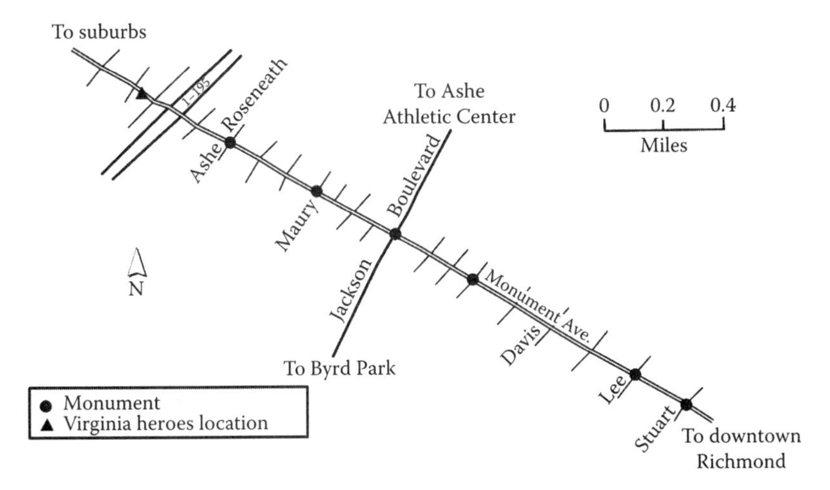

Figure 10.4 Map of statue locations along Monument Avenue.

the Civil War battles for Richmond. Thus, rather than being separated from the Confederate statues, the Ashe statue would be standing with the Confederate monuments inside their lines of defense against enemy forces. Richmond's African American city manager Robert Bobb, who had proposed the change, saw the move as being symbolic of a changed Richmond that had thrown off its segregationist past and embraced civil rights. He argued that the symbolism of the original proposal to place the statue west of Interstate 195 "would have been the statue was on the back of the bus. If all the other monuments were in the historic district, why not Ashe?"[47] He argued that placing the Ashe statue with the Confederate statues "will show the difference between what Monument Avenue was in the past to what it represents in the future."[48] He noted that the vote demonstrated that "Richmond is changing. We have changed. … It does reflect that we're a city for all the people of different views. … It's more than symbolic. It's real. This was a big day."[49] Thus, whereas placing the Ashe statue west of the highway would have been segregating it in a less historic part of the avenue, the approved location could be seen as integrating the avenue instead.

The commission's vote, however, led to a public outcry against the proposed location for the Ashe statue, setting off what Patricia Cornwell referred to as "another war."[50] Given the uproar over the decision, Richmond's city council scheduled a July 17, 1995, public hearing and vote to reconsider the Planning Commission decision, which helped set off a month-long debate over the proper location for the Ashe statue. As Richmond's African American mayor Leonidas Young noted, the debate was

> an important discussion about some of the most essential questions of our collective identity—the meaning of our traditions and symbols, the nature of heroism, the relation between our white and African-American populations and the function of public art in expressing the soul of a community.[51]

What Would Erecting a Statue of Arthur Ashe on Monument Avenue Mean?

The main question in the debate concerning whether Monument Avenue was the proper place for a statue of Arthur Ashe was, What was/is the meaning of Monument Avenue to Richmonders? The long-constructed meaning by Richmond whites was that as a tribute to the heroes of the Confederacy, Monument Avenue was a source of pride to Southerners. However, the Confederate monuments were an anathema to many Richmond blacks, who saw them as a symbol of their long-standing subjugation and who worked to make apparent that although the monuments may have been a source of pride to *white* Southerners, they were not necessarily a sense of pride to *all* Southerners (thereby making apparent the whiteness of this racialized landscape). Although the debate was in part about whether Ashe was worthy of being included on Monument Avenue, the debate was also important in that it allowed for a discussion of a question that had been long settled within (white) Richmond's public memory: Were the Confederate heroes immortalized on Monument Avenue really heroes? The fact that there could be a debate over the meaning of Monument Avenue was striking in that, unlike in the 1890s where African Americans had little political power in the city and could not meaningfully challenge the erection of the Lee monument, in the 1990s, in a majority-black city with a black mayor and black-majority city council, African Americans played an important role in shaping the debate over the meaning of Monument Avenue. Changes in the political power structure of the city by the 1990s allowed for narratives other than Monument Avenue as an avenue of heroes to be seriously considered. Although city councilwoman Shirley Harvey chastised Richmonders for being more concerned and vocal over where the Ashe statue should be located than such important issues facing Richmond as education, crime, and economic development, the vocal nature of this debate demonstrates the power that the Monument Avenue landscape (and its meanings) held over the city's citizens.[52]

A public opinion poll published in July 1995 in the city's largest newspaper, the *Richmond Times-Dispatch,* showed that Richmonders were opposed to the placement of the Ashe statue on Monument Avenue by a three-to-one margin, with a majority of both whites (by a four-to-one margin) and African Americans (by a three-to-two margin) polled in opposition.[53] However, although a majority of whites and African Americans opposed the placement of the Ashe statue on Monument Avenue, their reasons differed greatly and highlighted the different interpretations of the whiteness and symbolism of Monument Avenue. The following section examines three main issues concerning the meaning of the Monument Avenue landscape, and how they were approached in different ways by white and black opponents who wanted to keep the Ashe statue off Monument Avenue and by proponents (made up of both blacks and whites) who believed Monument Avenue was the proper place for the Ashe statue: First, was Monument Avenue's landscape a Confederate space? Second, did the Ashe statue deserve to be placed on Monument Avenue? (Or, alternatively, did Monument Avenue deserve the Ashe statue?) Third, what would placing the Ashe statue on Monument Avenue signify about Richmond?

Was Monument Avenue's Landscape a Confederate Space?

White opposition to placing the Ashe statue on Monument Avenue centered on several issues. Some felt that a modern statue of a late twentieth-century individual did not fit on a street built in the late nineteenth century and early twentieth century with statues commemorating events from more than a century ago. As one Monument Avenue resident argued, the Ashe statue "would be modern art in a turn-of-the-century, trolley-car neighborhood."[54]

For some whites, however, opposition to a Monument Avenue site for the Ashe statue was based on a conviction that, with its statues of Confederate heroes, Monument Avenue was a long-established sacred Confederate space and should remain as a site to venerate Confederate heroes and as a proud symbol of Southern history and heritage. (Though this raises the question of *whose* Southern history and heritage? Is Monument Avenue with its Confederate statues an inclusive proud symbol of all Southern history or an exclusive symbol of white Southern history?) As one opponent put it, "The statues on that street are dedicated to one cause and one single cause." Placing a statue of Ashe on the avenue would have "destroyed the significance of the street."[55] This objection to placing the Ashe statue on Monument Avenue was accompanied on more than one occasion by suggestions that if a monument to African Americans had to be placed on Monument Avenue, it should be of African American Confederate soldiers. As one white opponent put it, "If there's a hidden political agenda to put an Afro-American on Monument Avenue, they ought to honor the blacks who fought for the Confederacy" (though many historians question how many blacks willingly fought for the Confederacy).[56] As Leib argued, "Whiteness is apparent in this case, as for these Ashe opponents the black Southern experience is only important as it supports their white vision of

Southern history and social relations (blacks could only be worthy of veneration if they supported the white Confederate cause)."[57]

Although a majority of whites were opposed to the Ashe statue being placed on Monument Avenue, so too were a majority of Richmond's African Americans. One argument raised by black opponents agreed, in part, with white opponents, that being that Monument Avenue was Confederate space. However, rather than revering that space, black opponents argued that the landscape was to be abhorred and that a globally respected champion of human rights such as Ashe should not be forced to share space with those who fought to uphold slavery. Thus, as suggested by Ray Boone, the publisher and editor of the *Richmond Free Press* (at the time one of the city's leading African American newspapers), rather than a landscape to be honored, "Monument Avenue was constructed for a perverted reason of glorifying losers, the pro-slavery Confederate army in the Civil War."[58] Rather than celebrating heroes to be revered, Boone suggested, "Monument Avenue … celebrates the worst in our history—traitors who fought against human freedom."[59] Thus some white and black opponents of placing the Ashe statue on Monument Avenue agreed in their interpretation that the avenue was a "Confederate" space. However, although white opponents saw this as a landscape of heritage and pride, black opponents saw it as a landscape of hatred and loathing.

Black and white proponents of placing the Ashe statue on Monument Avenue rejected the idea that Monument Avenue was solely a Confederate space. To make this argument, proponents had to shift the dominant discourse that Monument Avenue was a place where only Confederates could be honored. The leading public supporter of placing the Ashe statue on Monument Avenue, former governor Wilder, tried to make this shift by first arguing that "Virginia's place to be recognized by Virginians is Monument Avenue," and then asking, "When you look to see Virginians who have made contributions and not see one single person of African descent, what does it say?"[60] Thus, Wilder argued that rather than being the exclusive province of Confederate heroes, all worthy Virginians should be considered for enshrinement on Monument Avenue (though, if that were the case, one could wonder where the likes of Patrick Henry and Thomas Jefferson were) and that, therefore, it was unconscionable that no African American Virginians had thus far been so honored.

Did the Ashe Statue Deserve to Be Placed on Monument Avenue? (or, Alternatively, Did Monument Avenue Deserve the Ashe Statue?)

Not all white opposition to placing the Ashe statue on Monument Avenue was explicitly based on the argument that, by virtue of it being a Confederate space, by definition Arthur Ashe (or any African American) could not be honored on Monument Avenue. Instead, some whites argued that as a tennis player, Ashe was not a hero on par with the Confederate generals because he did not risk his life to defend his nation and therefore his presence on the avenue would demean the Confederates' hero status. As one opponent put it about why Ashe should not be enshrined on Monument Avenue,

> True, he did a lot for tennis and was a nice person, and he inspired young people to make something good of their lives. Ashe's statue should be placed at the tennis courts. ... That would inspire more people to play tennis. ... There should not be a statue on Monument for any sports person, as I do not consider any of them heroes, unless they fought for our country. I always will think of Monument as showing our heritage to people from everywhere.[61]

Although this line of reasoning denied Ashe's humanitarian efforts (as one opponent put it, "Let's put this man in true perspective: Arthur Ashe was simply a tennis star who won Wimbledon one time"), it again raises the question of whose heroes and whose country the statues on Monument Avenue represent.[62] The assumptions and beliefs of Lost Cause Southern identity through the Jim Crow era resonate through this line of reasoning that white Southern heroes are, by definition, Southern heroes and that white Southern heritage, by definition, is Southern heritage. The black Southern experience has been largely invisible, both within (white) Southern memory and, by extension, on the Southern landscape.[63]

However, African American opposition to placing the Ashe statue on Monument Avenue turned this argument around, making the case instead that Ashe was too good a person and role model to be placed on Monument Avenue with those who fought to uphold slavery. Mayor Young, who opposed the Monument Avenue site, suggested, "Monument Avenue holds very harsh feelings for African-Americans."[64] He argued that Ashe would not be properly honored by having his statue put "in a place where it has been perceived that only white heroes go."[65] *Richmond Free Press* editor and publisher Ray Boone was also opposed for fear that by placing a hero such as Ashe on Monument Avenue, "you would be giving credence to the false proposition that these [Confederates] were heroes" as well.[66] Or, as one black opponent suggested, "Arthur Ashe doesn't belong with those racists. What Monument Avenue needs is a bulldozer."[67] As Leib pointed out, "Arguing that Ashe was too good to stand with the Confederates demonstrates an evaluation of the relative worthiness of both Ashe and the Confederates, in this case with Ashe being morally superior."[68] Unlike some whites, who argued that placing the Ashe statue on the avenue would demean the hero status of the Confederates, some blacks argued that placing the statue on the avenue would demean the hero status of Ashe.

If placing the Ashe statue on Monument Avenue would demean his status, then the question for African American opponents became one of where the monument should be placed if not on Monument Avenue. Black opponents of the Monument Avenue site had several suggestions. One was that the statue should go at the tennis courts at Richmond's Byrd Park, where Ashe was not allowed to play while he was growing up. The symbolism would be to show that Ashe and Richmond had overcome segregation. Although this suggestion was seriously considered by city leaders, it came in spite of the fact that Ashe's brother Johnnie argued against the Byrd Park site, stating that the site had no meaning to his brother, and that the statue's sculptor, Paul DiPasquale, said that Ashe had told him that he did not want to be remembered as just a tennis player.[69] City councilman Chuck Richardson, a strong proponent of the Monument Avenue site, objected to the Byrd Park tennis court location, especially when suggested by vocal whites, by arguing that placing the statue at a tennis court would reinforce the stereotype that "Blacks are good entertainers and sports people. Our talents go much further than that."[70] At the same time, one person was so eager to see the statue installed at Byrd Park and not Monument Avenue that shortly before the July city council meeting he offered to pay the entire amount necessary to complete the statue ($400,000) if DiPasquale would agree to place it at Byrd Park.[71]

Another suggestion for the Ashe statue was for it to be placed somewhere within one of Richmond's black neighborhoods, rather than placing it in the predominantly white Fan District neighborhood where Ashe would not have been welcomed as a child. As one black opponent of the Monument Avenue site put it, the statue should be put in the neighborhood where Ashe grew up "so that poor kids coming up could see where he came from. … Folks on Monument Avenue don't need to be encouraged."[72] For proponents who argued that Ashe transcended race, this argument that the statue should be placed only in a black neighborhood, thereby having a segregating effect rather than an integrative effect, angered them, again especially when it was made by white opponents. Black *Richmond Times-Dispatch* columnist Michael Paul Williams argued that the idea that the Ashe statue should be placed only in a black neighborhood was one of the "most disquieting aspects" of the debate:

> This thinking has no place in 1995, particularly regarding a figure as cosmopolitan and universally respected as Ashe. The segregation that afflicted Ashe during his lifetime has no place in discussions of how to treat his memorial. And besides: Who says black

> kids don't travel on Monument Avenue? And who says white kids can't benefit as much from the lessons of Ashe's life? The value he placed on education. His determination in the face of racism. The sportsmanship he displayed on the tennis court. His concern for the oppressed. His courage in the face of a deadly disease. Ashe should be a hero for all of Richmond, not just black people.[73]

If, as proponents of the Monument Avenue site had argued, "membership" on Monument Avenue was open to all Virginians, proponents then had to make the case that Ashe was worthy of inclusion. Although white opponents concentrated on Ashe's athletic ability, Ashe's proponents looked at his entire life work, emphasizing his accomplishments overcoming his childhood living under segregation, his human rights efforts, and his philanthropy both in Richmond and around the world. Thomas Chewning, a leading Richmond business executive and longtime friend of Ashe and cochair of the fund-raising effort for the statue, argued, "Ashe was a real moral leader ... one of the exceptional people of our time. ... What he stood for is what is best in all of us," and felt that Ashe should be mentioned "in the same paragraph" with Mohandas Gandhi and Martin Luther King Jr.[74] City councilman Richardson, while recognizing that Ashe was not a military leader as were the Confederates, nonetheless argued, "Arthur Ashe was a hero. He was a warrior to many people. He was a fighter. Arthur Ashe ... has already earned his right on Monument Avenue, whether we put him there or not."[75]

Although some white opponents were appalled at the thought that Ashe could share space with the Confederate leaders, Ashe's proponents were at least tacitly acknowledging that if Ashe's hero status made him eligible for inclusion on the avenue, then the Confederates had to be considered heroes too. Proponents tried to make a temporal distinction, suggesting that the Confederates were heroes of a previous time but that Richmond had changed dramatically so that Ashe could be considered a hero today. Former governor Wilder noted that the Confederates were "heroes from an era which would deny the aspirations of an Arthur Ashe. He would stand with them saying, 'I, too, speak for Virginia.'"[76] As Chewning argued in supporting Ashe's inclusion on Monument Avenue, "Putting his statue in line with Confederate leaders would show just how far Richmond had come without turning our back on our earlier history."[77] Thus proponents argued that because Monument Avenue was the place where Virginia's heroes were recognized, then that is where the Ashe statue should go. Indeed, as one speaker at the July city council meeting reasoned, in widely reported remarks, "If you put him in Byrd Park, years from now, people will say Arthur Ashe must have been a great tennis player. If you put him on Monument Avenue, people will say he was a great man."[78]

What Would Placing the Ashe Statue on Monument Avenue Signify about Richmond?

The final issue in the debate over the location of the Ashe statue was about what placing the statue on Monument Avenue would signify about Richmond. Although it is impossible to know how many felt this to be the case, there was an overtly racist tone to some of the debate, as some whites just did not want a black person memorialized on the same street with the Confederates. At the July city council meeting, several born and bred white Richmond speakers suggested that an Ashe statue placed on Monument Avenue would diminish their "quality of life." City councilman Larry Chavis voted for the Monument Avenue site because "I knew that my vote against Monument Avenue would condone that mentality."[79] Others were more vile in their interpretation of what placing the Ashe statue on Monument Avenue would signify. As one caller to a local Richmond radio station put it, "We need to protect our heritage. … We don't need blacks on Monument Avenue. … They've taken over our city; they've tried to take over our government. If you've got daughters like I've got daughters, they're trying to take them over too."[80]

For proponents, however, it was important for the Ashe monument to join the Confederates on the city's most famous street to send, in former governor Wilder's words, "a transcending message" that African Americans were now equals in the city.[81] Councilman Richardson suggested that an Ashe statue joining Lee, Jackson, Davis, Maury, and Stuart would "bring racial justice to Monument Avenue." He argued that Ashe's role in fighting for human rights and against apartheid made him the perfect person to integrate the avenue and symbolized the importance of civil rights to the city's history, suggesting the "hand-me-down ideals those individuals represent is the very thing that chased Arthur out of the city. The Civil War is part of our history. Now we have another part—civil rights. … For Arthur to take his final stand in the midst of what he has always fought, I think it would be fitting."[82]

The most eloquent supporter of Ashe's inclusion on Monument Avenue was *Richmond Times-Dispatch* columnist Michael Paul Williams. He argued it was finally time for white Richmonders to recognize that Monument Avenue was not a proud symbol for all Richmond residents:

> The time has come for people to acknowledge the anguish that Monument Avenue's symbolism has inflicted on black Richmonders. And to accept the fact that, for many of us, the strip is not hallowed ground but a painful reminder of black subjugation.

> White Richmonders must also accept the fact that in a predominantly black city, it is inappropriate to reserve a street for a cause many residents find offensive.[83]

Williams linked the battle to include the Ashe statue on Monument Avenue with the struggle by Richmond's blacks to end segregation, suggesting, "We've put a black man [Wilder] in the governor's mansion. But Monument Avenue is another matter." To Williams, erecting the Ashe statue on Monument Avenue would represent a major step toward ending the vestiges of the Jim Crow era:

> With checkered success, we've desegregated schools, governments, and businesses and neighborhoods. But the Bronze Club on Monument Avenue remains resistant to change. The sentiments may as well be etched in stone: "Whites Only." Arthur Ashe was stymied by a racist Richmond. Let's not plant his monument amid the shabby trappings of segregation, '90s style. We can't afford to repeat past mistakes. The world is watching.[84]

As Leib noted, proponents argued that placing the Ashe statue on Monument Avenue was an important symbol of what Richmond was in the past and where Richmond is today, because "given the powerful symbolism imbued in Monument Avenue, it was the only logical site for the statue and would send a strong message about the ending of racial divisions in Richmond."[85]

City Council Vote

Although it was unclear going into the city council meeting on July 17, 1995, whether the council would vote to place the Ashe statue on Monument Avenue, following a seven-hour debate with local, national, and international press in attendance, the council voted seven to zero with one abstention to erect a statue of Arthur Ashe on Monument Avenue at the intersection of Roseneath Road, three blocks west of the Maury statue (see Figure 10.4). Several city council members noted the larger significance of the vote. White councilman (and now Virginia's governor) Timothy Kaine suggested the vote would "symbolically open up Monument Avenue."[86] For Councilman Richardson, the vote had larger meaning in that it was "a symbolic move forward for Richmond and for America."[87]

Although the city council vote did not fully settle the issue, the vote paved the way for the Ashe statue to be placed in the intersection at Monument Avenue and Roseneath Road, at the western end of the monuments of Monument Avenue, where it sits today.[88] The statue was dedicated in 1996 and represents the first statue added to the avenue in sixty-seven years. At the unveiling ceremony in July 1996, former governor Wilder noted that Monument Avenue is now "an avenue for all people. ... Today, I feel more pride and relevance in being here on Monument Avenue than I have at any time in my life."[89]

This is not to suggest that Monument Avenue's landscape is now settled and that both proponents and opponents of the avenue's statues have not tried to alter the landscape's meaning for their own purposes. At various times over the past decade, Monument Avenue's Lee statue has been defaced. In 1998 the phrase "This is a monument to racism" was spray painted on the Lee monument, while in 2000 someone spray painted "Kill White Devil" on it.[90] In January 2004, two days before Robert E. Lee's birthday and the federal Martin Luther King, Jr. holiday, vandals defaced the Lee statue again, this time spray painting "Happy Birthday, MLK" on one side and "Death to Nazis" on the other.[91] As well, in 2000 black city councilman Sa'ad El-Amin led an unsuccessful campaign to end the city's annual appropriation to maintain the Confederate statues along Monument Avenue, arguing that the city should not be subsidizing "a tribute to slavery" (though El-Amin wanted the city to maintain Ashe's statue, noting that his "personal history isn't offensive to anyone").[92]

The other issues of race and Richmond's landscape have not been settled. For example, in 1999 a second major debate occurred over issues of race, the Civil War, and Richmond's symbolic landscape, as black city councilman Sa'ad El-Amin successfully attempted to remove a large head-and-shoulders mural of Lee from a historical display overlooking the centerpiece of Richmond's new riverfront Canal Walk redevelopment project. In the end, a compromise was reached, and a new mural of Lee was installed at the Canal Walk. However, during the debate about the mural, former Ku Klux Klan leader David Duke came to Richmond and protested the removal of Lee's mural at a news conference held in front of the Ashe statue on Monument Avenue.[93] Although less controversial, questions also have been raised in recent years about the city council's decision to rename two bridges in one of Richmond's black neighborhoods for local civil rights leaders (that had been previously named for Confederate generals) and about a private group's placement in 2003 of a statue of Abraham Lincoln near one of the city's Civil War sites.

Conclusions

The past century has seen the making and remaking of the physical and symbolic landscape of Monument Avenue. The making of this white racialized landscape in the late nineteenth century and early twentieth century was part of the larger effort to create the Lost Cause identity of the white

South and to impose white domination over the region's newly freed African American citizens. At the same time, however, African Americans did question the avenue's creation and meaning over time, thus challenging this Lost Cause version of the witting autobiography of white Richmond (and the white South). The status of Monument Avenue as a symbol of white power in Richmond remained firmly entrenched among whites and African Americans into the civil rights era, so much so that Richmond's white leadership took steps to ensure that the Confederate monuments would be protected in case African Americans won political control of the city and would seek to remove the Confederate statues as a concrete symbol of a shift in power.

As Kirk Savage suggested, "Public monuments do not arise as if by natural law to celebrate the deserving; they are built by people with sufficient power to marshal (or impose) public consent for their erection."[94] Although African Americans did raise challenges to the Monument Avenue landscape, they were powerless to change the landscape or alter its generally accepted meaning. By the 1990s, however, political power in Richmond had shifted so that black voices could shape the meaning of Monument Avenue, as demonstrated in the debate about the Arthur Ashe statue. Alternative narratives were raised and taken seriously over the meaning of Monument Avenue and its Confederate statues, rather than their being a proud symbol of white Southern identity only and being only within white Southern historical memory. On the one hand, some black opponents of placing the Ashe statue on Monument Avenue viewed the avenue as a place of shame rather than a place of pride and viewed the Confederates enshrined there as villains rather than heroes. For proponents, although the Confederate statues may have been symbolic of an earlier time when African Americans were denied the rights of citizenship, Arthur Ashe's human rights legacy would serve as a corrective to those enshrined. Either way, the 1995 debate, rather than being simply a debate over Ashe, provided for a reexamination over the meaning of Monument Avenue and the decentering of its long-standing position as a prime white racialized landscape. Given the changes in population composition and political power in Richmond in recent decades, the debate over Ashe and Monument Avenue both reflects and shapes the changing nature of race and power in Richmond.

Notes

1. Best-selling mystery writer and Richmond, Virginia, resident, Patricia Cornwell, *Southern Cross* (New York: Berkley Books, 1999), 187.

2. J.I. Leib, G.R. Webster, and R.H. Webster, "Rebel with a Cause? Iconography and Public Memory in the Southern United States," *Geojournal* 52 (2000): 303–10; J.I. Leib and G.R. Webster, "The Confederate Flag Debate in the American South: Theoretical and Conceptual Perspectives," in *Beyond the Color Line? Race and Community in the New Century,* ed. A. Willingham (New York: Brennan Center for Justice, New York University School of Law, 2002), 221–42; and G.R. Webster and J.I. Leib, "Whose South Is It Anyway? Race and the Confederate Flag in South Carolina," *Political Geography* 20 (2001): 271–99.
3. C.R. Wilson, *Judgment and Grace in Dixie* (Athens: University of Georgia Press, 1995), 20.
4. K. Edwards, E. Howard, and T. Prawl, *Monument Avenue: History and Architecture* (Washington, D.C.: Historic American Buildings Survey, U.S. Department of the Interior, National Park Service, Cultural Resources, 1992), xiii.
5. J.I. Leib, "Separate Times, Shared Spaces: Arthur Ashe, Monument Avenue, and the Politics of Richmond, Virginia's Symbolic Landscape," *Cultural Geographies* 9 (2002): 286–312.
6. J. Cigliano and S.B. Landau, eds., *The Grand American Avenue, 1850–1920* (San Francisco: Pomegranate ArtBooks, 1994).
7. J. Busbee, "Richmond Rallies for Arthur Ashe," *Reckon, the Magazine of Southern Culture,* Winter 1996, 16–17.
8. R.G. Wilson, "Monument Avenue: Richmond, Virginia," in *The Grand American Avenue: 1850–1920,* 259–79 (esp. p. 259).
9. P.F. Lewis, "Axioms for Reading the Landscape: Some Guides to the American Scene," in *The Interpretation of Ordinary Landscapes: Geographical Essays,* ed. D.W. Meinig (New York: Oxford University Press, 1979), 11–32 (esp. p. 12).
10. R.H. Schein, "Teaching 'Race' and the Cultural Landscape," *Journal of Geography* 98 (1999): 188–90 (esp. p. 189).
11. C.R. Wilson, *Baptized in Blood: The Religion of the Lost Cause* (Athens: University of Georgia Press, 1980); G. Foster, *Ghosts of the Confederacy: Defeat, the Lost Cause, and the Emergence of the New South* (New York: Oxford University Press, 1987); and D. Delaney, *Race, Place and the Law* (Austin: University of Texas Press, 1998).
12. Wilson, "Monument Avenue," 278.
13. J.M. McPherson, *Drawn with the Sword: Reflections on the American Civil War* (New York: Oxford University Press, 1996), 151.
14. J.I. Leib, "Robert E. Lee, 'Race,' Representation and Redevelopment along Richmond, Virginia's Canal Walk," *Southeastern Geographer* 44 (2004): 236–62 (esp. p. 239); and T.L. Connelly, *The Marble Man: Robert E. Lee and His Image in American Society* (Baton Rouge: Louisiana State University Press, 1977), 3.
15. J. DuPriest Jr. and D. Tice Jr., *Monument and Boulevard: Richmond's Grand Avenues* (Richmond, VA: Richmond Discoveries Publication, 1996), 8.

16. M. Tyler-McGraw, *At the Falls: Richmond, Virginia and Its People* (Chapel Hill: University of North Carolina Press, 1994), 209.
17. Wilson, *Baptized in Blood,* 29.
18. S.S. Driggs, R.G. Wilson, and R.P. Winthrop, *Richmond's Monument Avenue* (Chapel Hill: University of North Carolina Press, 2001), 53.
19. Tyler-McGraw, *At the Falls,* 209.
20. Driggs, Wilson, and Winthrop, *Richmond's Monument Avenue,* 53.
21. A. Ashe and A. Rampersad, *Days of Grace: A Memoir* (New York: Ballantine Books, 1993), 317.
22. Edwards, Howard, and Prawl, *Monument Avenue;* and Tyler-McGraw, *At the Falls.*
23. *New York Times,* "Richmond Street Stirs Up a Row," December 12, 1965, 89; Tyler-McGraw, *At the Falls.*
24. Driggs, Wilson, and Winthrop, *Richmond's Monument Avenue,* 236, 240.
25. Edwards, Howard, and Prawl, *Monument Avenue,* 99.
26. J.V. Moeser and R.M. Dennis, *The Politics of Annexation: Oligarchic Power in a Southern City* (Cambridge: Schenkman, 1982); C. Silver and J.V. Moeser, *The Separate City: Black Communities in the Urban South, 1940–1968* (Lexington: University of Kentucky Press, 1995); and L.A. Randolph and G.T. Tate, *Rights for a Season: The Politics of Race, Class, and Gender in Richmond, Virginia* (Knoxville: University of Tennessee Press, 2003).
27. K. Savage, "The Politics of Memory: Black Emancipation and the Civil War Monument," in *Commemorations: The Politics of National Identity,* ed. J.R. Gillis (Princeton, NJ: Princeton University Press, 1994), 127–49 (esp. p. 134).
28. Moeser and Dennis, *The Politics of Annexation,* 77.
29. Quoted in ibid., 77–78.
30. M. Barone and G. Ujifusa, *The Almanac of American Politics, 1998* (Washington, D.C.: National Journal, 1997).
31. Edwards, Howard, and Prawl, *Monument Avenue,* 100.
32. J. Tooley, "Boulevard of Broken Dreams," *U.S. News and World Report,* November 11, 1991, 24.
33. Busbee, "Richmond Rallies for Arthur Ashe," 16.
34. Barone and Ujifusa, *The Almanac of American Politics,* 1458; T.N. Chewing, "Everything I Needed to Know about Business Ethics I Learned from My Parents, My Boss, and Arthur Ashe" (remarks to Beta Gamma Sigma, McIntire School of Commerce, University of Virginia, March 22, 2005), www.dom.com/about/speeches/032205_print.jsp (accessed May 22, 2005).
35. Ashe and Rampersad, *Days of Grace;* K. Moore, "The Eternal Example," *Sports Illustrated,* December 21, 1992, 19–26.
36. Moore, "The Eternal Example."
37. A. Ashe, *Off the Court* (New York: New American Library, 1981), 52.
38. Ashe and Rampersad, *Days of Grace,* 123.
39. Ibid., 141.
40. For example, M. Allen, "Just Plain Better Than Most of Us," *Richmond Times-Dispatch,* February 11, 1993, A1, A14.

41. Busbee, "Richmond Rallies for Arthur Ashe"; Driggs, Wilson, and Winthrop, *Richmond's Monument Avenue.*
42. Quoted in Chewing, "Everything I Needed to Know."
43. Quoted in Driggs, Wilson, and Winthrop, *Richmond's Monument Avenue,* 92.
44. J. Strauss, "Monumental Task," *Atlanta Journal-Constitution,* January 8, 1995, M1.
45. Edwards, Howard, and Prawl, *Monument Avenue,* 216.
46. Leib, "Separate Times," 296.
47. Quoted in G. Hickey, "Statue's Path Wasn't Smooth," *Richmond Times-Dispatch,* July 8, 1996, A3.
48. Quoted in G. Hickey, "City OKs Monument Site for Ashe," *Richmond Times-Dispatch,* June 20, 1995, A5.
49. P. Baker, "Landmark Decision in Richmond," *Washington Post,* June 20, 1995, B7.
50. Cornwell, *Southern Cross,* 187; M. Allen, "Return to Square One," *Richmond Times-Dispatch,* June 28, 1995, A1, A6; *New York Times,* "Richmond Hesitates on Tennis Star's Statue," June 27, 1995, B7.
51. Quoted in T. Campbell and M. Allen, "Ashe Family Big Factor in Outcome," *Richmond Times-Dispatch,* July 19, 1995, A10.
52. J. Roop, "The World Watched," *Richmond State,* July 20–26, 1995, 1–2.
53. M. Allen, "Many Oppose Monument Site," *Richmond Times-Dispatch,* July 19, 1995, A8.
54. Quoted in M. Allen, "Monument Site for Ashe Put in Doubt," *Richmond Times-Dispatch,* June 27, 1995, A1, A5.
55. Quoted in Hickey, "City OKs Monument Site for Ashe."
56. Quoted in *New York Times,* "Race-Tinged Furor Stalls Arthur Ashe Memorial," July 9, 1995, 11.
57. Quoted in Leib, "Separate Times," 299.
58. Quoted in *Richmond Free Press,* "City Council to Hear People on Memorial," June 29–July 1, 1995, A5.
59. Quoted in the *Economist,* "Advantage Who?" July 15, 1995, 20.
60. Quoted in Strauss, "Monumental Task."
61. P. Moody, letter to the editor, *Richmond Times-Dispatch,* July 1, 1995, A8.
62. Quote from P. Murphy, letter to the editor, *Richmond Times-Dispatch,* July 17, 1999.
63. Leib, "Separate Times"; Leib, "Robert E. Lee."
64. Quoted in M. Edds, "Clashing Views on Ashe Memorial," *Richmond Free Press,* July 13–15, 1995, 4.
65. Quoted in Allen, "Monument Site for Ashe Put in Doubt," A1.
66. Quoted in Edds, "Clashing Views on Ashe Memorial," 4.
67. Quoted in M. Allen, "Mayor Offers a Compromise," *Richmond Times-Dispatch,* July 17, 1995, A7.
68. Leib, "Separate Times," 300.
69. G. Hickey, "Offer Made to Pay for Ashe Statue in Byrd Park," *Richmond Times-Dispatch,* July 29, 1995, B1.

70. Quoted in T. White, "Symbolism, History Collide in Debate," *Richmond Afro-American,* July 13–19, 1995, A3.
71. Hickey, "Offer made to pay for Ashe Statue in Byrd Park."
72. Quoted in *New York Times,* "On Street Where Confederates Reign, Arthur Ashe May Too," June 18, 1995, A16.
73. M.P. Williams, "A Monument to Richmond's Strife," *Richmond Times-Dispatch,* July 3, 1995, B6.
74. Quoted in Edds, "Clashing Views on Ashe Memorial," 4.
75. M. Allen, "Ashe Family Has Diverse Views on Statue," *Richmond Times-Dispatch,* June 19, 1995, A4.
76. Quoted in *New York Times,* "On Street Where Confederates Reign, Arthur Ashe May Too."
77. Chewning, "Everything I Needed to Know."
78. Quoted in M.P. Williams and T. Campbell, "Ashe Family Big Factor in Outcome," *Richmond Times-Dispatch,* July 19, 1995, A8.
79. Quoted in G. Hickey, "Ashe Vote Eluded Mayor," *Richmond Times-Dispatch,* July 19, 1995, A8.
80. Quoted in P. Baker, "Richmond to Honor Ashe Alongside Rebel Heroes," *Washington Post,* July 18, 1995, A8.
81. Quoted in M.P. Williams, "Arthur Ashe Deserves Place on Monument," *Richmond Times-Dispatch,* December 12, 1994, B1.
82. Quoted in Allen, "Monument Site for Ashe Put in Doubt."
83. Williams, "A Monument to Richmond's Strife."
84. Williams, "Arthur Ashe Deserves Place on Monument."
85. Leib, "Separate Times," 302.
86. Quoted in *Richmond Free Press,* "Ashe Family Credited for Winning Vote," July 20–22, 1995, A5.
87. Quoted in G. Hickey, "Ashe Statue Will Go on Monument," *Richmond Times-Dispatch,* July 18, 1995, A1.
88. Leib, "Separate Times," 305–306.
89. M.P. Williams, "At Least Some of City's Ghosts Are Exorcized at Ceremony," *Richmond Times-Dispatch,* July 11, 1996, A1.
90. *Richmond.com,* "Lee Monument Defaced," 2000, www.richmond.com/printer.cfm?article=572938 (accessed May 23, 2005).
91. Leib, "Robert E. Lee."
92. Quoted in C. Johnson, "El-Amin: Stop City Funding for Confederate Statues," *Richmond Times-Dispatch,* May 10, 2000.
93. Leib, "Robert E. Lee."
94. Savage, "The Politics of Memory," 135.

CHAPTER 11

Naming Streets for Martin Luther King Jr.: No Easy Road

DEREK H. ALDERMAN

A Street Fight in Chattanooga

In 1981 Reverend M.T. Billingsley asked the city commission of Chattanooga, Tennessee, to name a street after slain civil rights leader Martin Luther King Jr.[1] Rev. Billingsley's request, which came on behalf of the Ministers Union he helped lead, took place just five days after King's birthday, which had not yet been made into a federal holiday, and less than a year after a controversial court verdict had incited civil disturbances in the city. An all-white jury had acquitted two of three defendants arrested in connection with the shooting of four black women. Ninth Street had been the location of the Ku Klux Klan-related shooting, and, not coincidentally, it was also the road that black leaders sought to rename for King. In addition, Ninth Street had long served as the city's black business district and was the site of a federally funded downtown redevelopment program.[2]

The street-naming request sparked several months of intense debate. One of the major opponents to the name change was a T.A. Lupton, a white real estate developer who owned a downtown office building on the west end of the street and was in the process of building another one there. Lupton argued that he might not be able to rent office space in a building with a King address because of the "racial overtones" it might create. He supported the renaming of East Ninth Street but not West Ninth Street, implying that the civil rights leader's memory would somehow be "out of place" there. He was quoted as saying, "West Ninth Street is not related to Dr. King. … [It] is no longer a solid black street. … It is no longer a residential street or rundown business street. It is a top class business street that can play a great part in the future of Chattanooga."[3] The developer went so far as to suggest that he would abandon or drastically alter his construction plans in the event that West Ninth was renamed.

Chattanooga's city commissioners acquiesced to pressure from Lupton and other opponents, refusing to rename Ninth. As a compromise, the commission offered to establish a plaza in King's memory. Street-naming proponents quickly dismissed this alternative and responded by organizing a march along Ninth Street in late April 1981. Armed with ladders and singing the civil rights anthem "We Shall Overcome," more than three hundred African Americans defiantly—albeit temporarily—renamed the street by pasting street signs and utility poles with green bumper stickers that read "Dr. ML King Jr. Blvd."[4] After this protest and an emotional request from a coalition of white and black ministers, the Chattanooga city commission reversed itself in July and agreed to rename all of Ninth Street for the civil rights leader as of January 1982.[5]

Controversy over the street renaming continued even after the city commission's decision. National Association for the Advancement of Colored People (NAACP) leaders would later encourage the boycott of a prominent Chattanooga hotel that had changed its mailing address from West Ninth to a bordering street, presumably to avoid being identified with Martin Luther King.[6] In the end, the street renaming did not stop Lupton, and he went forward with construction plans; however, the resulting office tower and its companion building, the corporate headquarters of the Krystal hamburger chain, do not have a King street mailing address. Instead, a private drive was created and the buildings reside on—presumably with no irony intended—Union Square.[7]

The street fight in Chattanooga was not an isolated event but part of a growing landscape movement in America. In fact, when petitioning to rename Ninth Street, several black leaders in Chattanooga cited the fact that cities such as Atlanta and Chicago and even other cities in Tennessee had already honored King with a street-name change. More than 730 cities and towns in thirty-nine states and the District of Columbia had a street named for King by 2003 (see Figure 11.1).[8] Commemorating King through street naming displays a strong regional concentration even as it is a national trend. Seventy percent of places with King streets are located

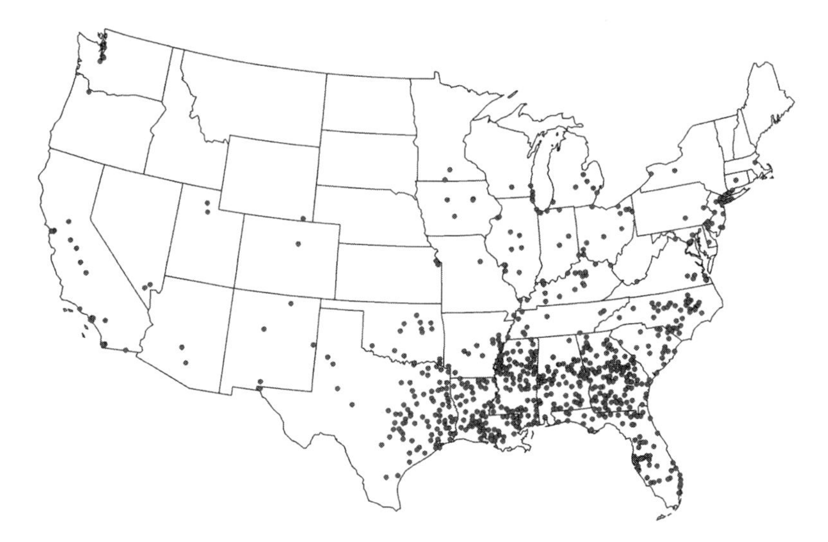

Figure 11.1 Distribution of streets named for Martin Luther King Jr., 2003. Source: Compiled by the author, Matthew Mitchelson, and Chris McPhilamy.

in the seven southern states of Georgia, Mississippi, Texas, Florida, Louisiana, Alabama, and North Carolina. Georgia—King's home state—leads the country with 105 named streets. Naming streets for King occurs throughout the urban hierarchy, from large cities such as New York, Los Angeles, and Houston to some of the country's smallest places such as Cuba, Alabama (population 363), Pawley's Island, South Carolina (population 138), and Denton, Georgia (population 269). Although street-naming struggles in metropolitan areas such as Chattanooga typically receive more publicity, it is worth noting that more than 50 percent of U.S. streets named for King are in places with a population of fewer than ten thousand people (see Figure 11.2).

Naming streets for King is widespread and often has been controversial. The practice and its controversies provide insight into the intersection of race and landscape in the United States. King-named streets reflect the increased cultural and political power of blacks and the liberalization of white attitudes even as they also are sites of struggle for African Americans. I have suggested in previous work that these streets serve as memorial arenas—public spaces for interpreting King's historical legacy and debating the connotations and consequences of commemorating him.[9] As evident in the Tennessee case, there can be significant differences in the extent to which people personally identify with King and wish to have their street associated with him and, as they perceive it, the black community. Despite the victory in Chattanooga, African Americans in many

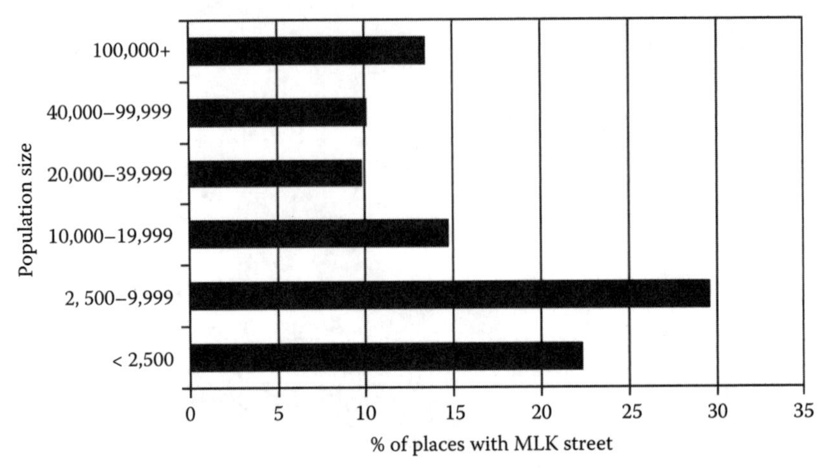

Figure 11.2 Distribution of streets named for Martin Luther King Jr. by city population.

other cities have been unsuccessful in renaming thoroughfares that cut across business districts and connect different racial groups. King's name is frequently attached to minor streets or portions of roads located entirely within poor, black areas of cities (see Figure 11.3).[10] This has led, in turn, to the widespread belief that all King streets are this way, even though

Figure 11.3 Martin Luther King Circle in Phoenix, Arizona. The small cul-de-sac is invisible on some city maps. Until recently, the six homes on the street were owned exclusively by African Americans. Past attempts to rename a major road have met with resistance in a state that refused to establish a paid holiday in King's honor until it lost hundreds of millions of dollars from convention cancellations and the National Football League's decision to host the 1993 Super Bowl elsewhere. (Photograph by Keli Dailey, reproduced with permission.)

there are prominent roads bearing his name. As journalist Jonathan Tilove so keenly observed, "It has become a commonplace of popular culture to identify a Martin Luther King street as a generic marker of black space and not incidentally, of ruin, as a sad signpost of danger, failure, and decline, and a rueful rebuke of a people's preoccupation with symbolic victories over actual progress."[11] However, Tilove has found that King streets are important centers of black identity and community within America. Rather than a hollow gesture, street naming for many black activists is about gauging society's relative progress in fulfilling the goals of the civil rights movement. For instance, when marching down Ninth Street in Chattanooga, black leaders characterized their street-naming struggle as an opportunity not only to celebrate King's achievements but also "to test whether 'equality' and 'justice' for all are valid statements, or whether they have no meaning at all."[12]

This chapter introduces King street-naming practice and explains why it is important and controversial. My intent is to identify (1) the political origins and historical development of the street-naming movement, (2) the symbolic qualities of street naming as a means of commemorating King, and (3) the political controversy and struggle that underlies street-naming practices. Streets named after King illustrate the important yet contentious ways in which race, place, and memory intersect through the American landscape. They provoke "fervent debate about the meaning of his [King's] life and what kind of street would do him credit," revealing important divisions between blacks and whites as well as social contests within African American communities.[13]

Origins of Streets Named after Martin Luther King

The movement to name streets after Martin Luther King originated squarely within black community activism. King's commemoration is part of an ongoing effort on the part of African Americans to address the exclusion of their experiences and achievements from the national historical consciousness. According to Joseph Tilden Rhea, this movement goes beyond the country's general embrace of multiculturalism. Rather, African Americans and other racial and ethnic groups are using direct political action to challenge and change the commemoration of the past within cultural landscapes, constituting what Rhea called the "race pride movement."[14] The race pride movement has had an impact on not just street-naming patterns but also other commemorative forms such as statues, museums, preserved sites, heritage trails, and festivals. Although less ornate or ostentatious than these memorials, street naming has become one of the most common and visible strategies for African Americans to elevate public recognition of King as well as a host of other figures identified with the struggle for equal rights, such as Rosa Parks, Thurgood Marshall, Malcolm X, and Harriet Tubman. As author Melvin Dixon observed, "Not only do these [street] names celebrate and commemorate great figures in black culture, they provoke our active participation in that history. What was important yesterday becomes a landmark today."[15]

In the case of street naming, African Americans are not simply honoring the historical achievements of a single individual but seeking to establish the public legitimacy of all blacks. Commemorating King is inseparable from a broader consideration of racism and race relations, especially a desire to reverse the control historically exercised by whites over racial and ethnic minorities. Black activists envision being able to engage in commemoration as part of the democratization of society and the gaining of a greater political voice. For example, after Chattanooga's city commission finally approved the renaming of Ninth Street for King, NAACP leader George Key concluded that the decision shows that "black citizens are full citizens of Chattanooga and have a right to be considered in what goes on in [the city]."[16] As Karen Till argued, street-naming struggles such as seen in Tennessee "often reflect larger social (power) disputes about who has authority to create, define, interpret, and represent collective pasts through place."[17]

Given the central role that black activists play in initiating the street-naming process, it is not surprising that a strong relationship exists between the likelihood of a city or town identifying a street with King and the relative size of its African American population (see Figure 11.4).

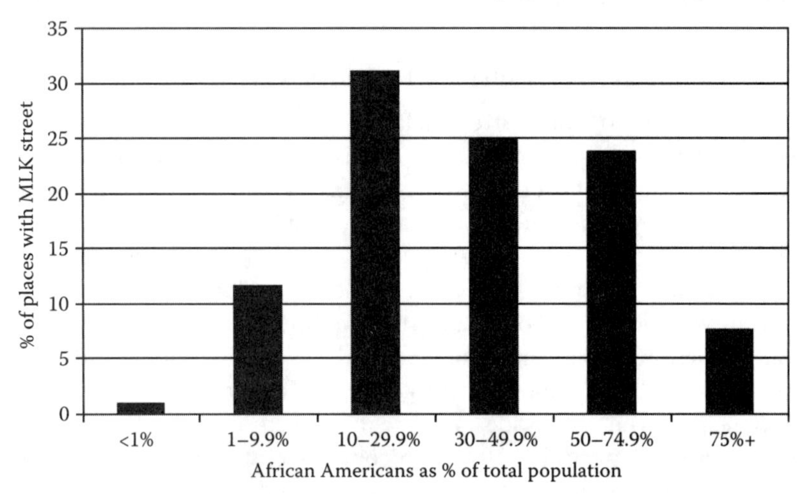

Figure 11.4 Distribution of streets named for Martin Luther King Jr. by relative size of city's African American population.

On average, African Americans constitute approximately 37 percent of the population in a location with a street named for King. More than a third of the time, African Americans make up 50 percent or more of the population in places with a street named after King. Street-naming campaigns are often conducted by local chapters of the NAACP, the Southern Christian Leadership Conference, an organization that King once led, and various other black-led community improvement associations and coalitions. The church is often an important participant in the naming process, as it has been in African American culture in general. Churches are one of the nonresidential establishments most frequently found on streets that bear King's name.[18] In Chattanooga, it was a union of black ministers who spearheaded the renaming of Ninth Street. Ultimately it was this group's ability to form a coalition with white clergy in the city that helped sway the opinion of local elected officials. In the case of the small town of Metter, Georgia, a local black pastor led the movement to rename a street. The unveiling of Martin Luther King Boulevard took place on the Sunday before the 1996 King holiday, and the dedication service began and ended with prayer and the singing of church hymns. During the service, those in attendance read a litany of dedication, pledging themselves to the ideals of peace, freedom, and equality.

The race pride movement and the commemoration of King are relatively recent developments; however, they are not entirely new and, in some way, signal a return to an earlier American tradition. The United States has a long history of honoring patriot heroes and using commemorative symbols such as monuments, museums, and place names to focus public attention and identification with certain political values and visions of history. For instance, Zelinsky found that 25 percent of counties and 10 percent of streets in the United States are named after national notables or carry other patriotic references.[19] Although the growing movement to memorialize King and other civil rights leaders represents a return to honoring inspirational heroes, we should not forget that earlier patterns of commemoration were almost entirely devoted to honoring white historical figures such as presidents and the country's founding fathers. Streets named for King challenge the country's dominant historical memory in that they ask citizens to view the past and its heroes in much more diverse terms, ones that specifically address experiences common to being black in America. In their book *Presence of the Past,* historians Roy Rosenzweig and David Thelen found significant racial differences when surveying Americans about how they value and identify with the past. Perhaps it is not surprising that African American respondents are much more likely than whites to cite the assassination of King as an event in the past that has most affected them.[20]

The contributions of African Americans certainly do not begin or end with King, but he has become the most widely identified symbol of the civil rights movement and black heritage in general, sometimes at the historical neglect of lesser known activists, including women.[21] King memorializing began after his assassination in 1968, but such efforts did not receive immediate widespread approval.[22] Four months after the civil rights leader's death, the city of Chicago, Illinois, renamed South Park Way, perhaps making it the country's first street named after King. Although the road stretches for several miles, it does not leave the city's predominantly African American South Side. On South Park Way, an African American church chose to use one of its side streets as an address rather than be identified with the civil rights leader. The pastor at the time, Reverend Joseph Jackson, was a conservative opponent of the civil disobedience campaign and a "bitter rival of King's in national black Baptist circles."[23] In 1972 officials in Montgomery, Alabama, approved and then quickly rescinded a measure to rename a street for King. The White Citizens Council—an organization that King battled during the city's famous bus boycott—opposed the name change because white-owned businesses and a white Masonic lodge were found on the street and the street was not located entirely in the black community.[24] It was not until 1976 that Atlanta, Georgia—King's hometown—placed his name on a street. Commercial interests opposed the naming of a street in his honor, though without success.[25] Today, Martin Luther King Jr. Drive, on the west side of the city, is the location of significant economic development and a major landmark in the city's tourism industry.

King's status rose when the federal government established a holiday to honor him in 1983, although passage of the holiday came fifteen years after first being proposed in Congress and after debate among black leaders about the most appropriate date to observe.[26] There are indications that the King holiday has helped propel the street-naming movement. It is difficult to know exactly when many of the country's cities and towns named a street for King; however, a preliminary survey of Georgia municipalities revealed that only 13 percent of responding communities named a street before the King federal holiday was signed into law by Ronald Reagan. More than a third of responding cities said they had renamed a street between 1984 and 1989. The remaining 52 percent of street naming in Georgia occurred after 1990.[27] In many instances, local King holiday celebration commissions organize street-naming campaigns. Many petitions to name a street are brought before local governments immediately before or after the King holiday in January. Although the holiday made King an officially recognized icon and gave further legitimacy to commemorating him, it was perhaps what the holiday could not offer African Americans that most inspired their requests to rename streets. Unlike the holiday, which comes just once a year, a commemorative street name provides a physically permanent memorial that is present all the time. The King holiday remains controversial. Some local governments still refuse to recognize it, and only 26 percent of the country's businesses give a paid day off to their employees.[28] Consequently, naming a street represents a more concrete way for communities to display their commitment to King's memory and ideals. African Americans are also pursuing street naming because of its symbolic qualities.

Symbolism of Streets Named after Martin Luther King

The seemingly mundane practice of street naming invokes intense emotional response. The attraction of commemorative street naming to African American communities is multifaceted, and commemorating Martin Luther King in the urban landscape transcends an immediate concern with simply naming roads to symbolically mediate myriad questions of race and racism in American life. King street-naming practices mark concerns for and debates about political meaning, power and resistance, historical representation, social justice, public space and infrastructure access, urban diversity, and community memory and identity. These debates extend beyond African American communities, to open up often long-standing American cultural, political, social, and economic tensions.

As Maoz Azaryahu suggested, "Street names are more than a means of facilitating spatial orientation. Often they are laden with political meanings and represent a certain theory of the world that is associated with and supportive of the hegemonic socio-political order."[29] In other words, street names, along with other forms of memorialization, participate in legitimizing a selective vision of the past, making historical representations appear to be the natural order of things. Places named after historical people or events are important symbols within a country's political culture and often are manipulated by state leaders or elites to reconstruct national identity.[30]

The power of street naming means that it also can be used by historically subordinate or marginalized groups as a form of resistance to challenge prevailing ideas about their identity and importance within society.[31] Black activists often are aware of the counterhegemonic potential of naming streets for King. A street-name change symbolizes a shift in the racialized balance of power between whites and blacks as well as racial progress within communities. Such feelings were expressed in Statesboro, Georgia, in 1994 when citizens were asked to submit their views on whether a major perimeter highway should be named: "I want to convey my personal suggestion to name the road Martin Luther King Parkway. I agree that this naming would affirm an important segment of our ... [city] and county population and would be a healing and unifying act."[32] Some black activists in Statesboro conceptualized naming the perimeter for King as an ideological weapon against an oppressive white cultural structure. For instance, the local NAACP leader argued that several of the city's existing streets were named after racist whites and that local leaders had never asked members of the black community if these streets bothered them. He added, "We need a street that honors a man that symbolizes something different about race. Dr. King stood for equality."[33] In this respect, the naming of streets is conceptualized by some blacks as an antiracist practice, a way of inscribing a new vision of race relations into the American landscape.

African Americans have sought access to several mediums of public commemoration, but street naming has proved to be especially important. Symbolically, street naming is the latest chapter in a long line of African American struggles for social justice in the area of mobility and transportation. Streets do not operate in a race- and class-neutral society; a politics underlies their organization, use, and meaning.[34] Although transportation racism is usually understood in terms of the inequity found in highway spending, road improvements, and mass transit planning, it also can refer to the barriers that confront racial and ethnic minorities as they seek to define the symbolic identity and meaning of roads as public spaces. From the Underground Railroad to the Freedom Riders, black communities have long looked to movement and transportation as conduits for challenging and changing the racial order. Sally Greene—a white city councilor in Chapel Hill, North Carolina, and a supporter of renaming the city's Airport Road for King—expressed the strong connection she saw between street naming and the larger history of African American struggles for equality:

> Under Jim Crow laws, blacks had a hard time just making a road trip. They had to pack their own food, even their own toilet paper, for they didn't know if they would find a restaurant that would serve them or even a gas station where they could use the bathroom. ... Then came Dr. King and the bus boycott and the push for the public accommodation law. ... Mobility, the freedom to travel the public roads without fear and with assurance that you get what you needed—these were the basic goals for King. Thus, I can't think of a better way to honor Dr. King than with a road naming.[35]

Aside from the specific historical experiences of African Americans, commemorative street naming is, in general, an important vehicle for bringing the past into the present. The seemingly ordinary and practical nature of street names makes the past tangible and intimately familiar. Because of its practical importance, street naming inscribes its ideological message into many practices and texts of everyday life.[36] Yvonne Aikens, who pushed to have a street named for King in Tampa, expressed this point well:

> A street touches more people than if they had just named a building after him downtown. ... People who wouldn't go to a building or a park named for King drive on a major thoroughfare such as Buffalo (now Martin Luther King, Jr. Blvd.) for business or personal reasons. ... They see the name at intersections, on signs pointing to the road, on maps. It pops up on addresses, letters, business cards, constantly keeping King's name before the public. ... More people come in contact with it.[37]

By all indications, it appears that Tampa's King Boulevard is a highly visible point of contact for the public. It extends for more than fourteen miles, connects with two interstate highways, and serves as an address for more than 550 nonresidential or business establishments.[38]

The power and politics of commemorative street naming lie in its dual and simultaneous existence as historical referent and form of spatial identification. A street name's practical nature does not necessarily lessen its symbolic function. Rather, the commemorative importance of street names comes from their status as markers of location. Public commemoration is not simply about determining the appropriateness of remembering the past in a certain way but a struggle over where best to place that memory within the cultural landscape.[39]

The symbolism of location in commemorating King was perhaps no more apparent than in Brent, Alabama, when blacks protested attaching King's name onto a road leading to a garbage dump. Reverend W.B. Dickerson petitioned the city council to rename another, more prominent street, and said, "We want [Martin Luther King Street] up where people can really see it."[40] Similarly, in March 2002 African American activist Torrey Dixon petitioned the city council of Danville, Virginia, to rename Central Boulevard, a major commercial thoroughfare. Though unsuccessful, he considered the boulevard an "appropriate street" because its central location and high volume of traffic would ensure that many people would see King's name. Dixon even refused to rename an alternative street that had a strong historical connection with King's visit to Danville in 1963 because he claimed it was a "low class neighborhood."[41] These situations affirm Nuala Johnson's suggestion that location is not simply the "incidental material backdrop" for memory but plays an active role in constructing the meaning of commemoration.[42] As Jonathan Tilove observed, "To name any street for King is to invite an accounting of how the street makes good on King's promise or mocks it."[43]

For many African Americans, streets have a geographic connectivity that contributes to their symbolism. This was made clear in one editorial: "Renaming a street is a uniquely appropriate way to honor King. Streets unite diverse neighborhoods. They touch all ages, all races, all economic levels, and the resident and the visitor equally. They link people and places that otherwise would remain insular."[44] The notion of connectivity is particularly relevant to commemorating King. For instance, African American activist Allen Stucks envisioned Martin Luther King Boulevard in Tallahassee, Florida, in terms of King's goal of racial integration: "Rev. King was about togetherness. ... If his name was going to be on a street in Tallahassee, it had to be on one that connected one neighborhood to another. And it had to be one you could find without having to wiggle through the black community." In the case of King Boulevard in Tallahassee, the street "connects one of the nation's oldest historically black universities to the entire city. It traces through black neighborhoods, white neighborhoods, businesses, parks and cemeteries."[45] The street named for King in Austin, Texas, also crosses racial lines, the result of the passionate yet fatal lobbying of J.J. Seabrook. He died of a heart attack while pleading with the city council not to restrict the named street to the black community.[46] Street naming is a potentially powerful form of commemoration because of its capacity to make certain visions of the past accessible to a wide range of social groups. However, it is this potential to touch and connect disparate groups—some of which may not identify with King—that also makes street naming controversial.

The symbolic meaning of streets named after King is also about how they connect with a larger memorial landscape, including other named places, historical markers, murals, and monuments. Found along Martin Luther King Jr. Drive in Asheville, North Carolina, is a recreational park that also bears the civil rights leader's name. The centerpiece of the park is a life-size statue of King leading two small children. It is not uncommon for King to be remembered alongside other historical figures (see Figure 11.5). Streets named after King and Malcolm X intersect in Dallas, Texas (as well as in Harlem, New York), creating an interesting moment for reflecting on the similarities and differences in how these two leaders worked for civil rights. At the Martin Luther King Memorial Gardens in Raleigh, North

Figure 11.5 Freedom Corner Monument at the intersection of Martin Luther King and Medgar Evers boulevards in Jackson, Mississippi. Streets named for King do not exist in a symbolic vacuum but become connected to other memorial forms, other historical figures, and other political causes. In 2001, for example, a group led by a black city councilman burned a state flag at Freedom Corner in calling for removal of the Confederate battle emblem from Mississippi's official banner. (Photograph by Elizabeth Hines, reproduced with permission.)

Carolina, a bronze statue of King overlooks a major road bearing his name. Next to the statue is a marble fountain inscribed with the names of civil rights leaders from the Raleigh area, creating a place where visitors can interpret the interweaving of national and local civil rights movements. Yet bringing national and local civil rights histories together can create contradictory landscape formations. For instance, the Ralph Mark Gilbert Civil Rights Museum in Savannah, Georgia, is located on Martin Luther

King Jr. Boulevard even though African American leaders—including Gilbert—tried to bar King from preaching in the city in the 1960s. They feared that the civil rights leader's presence would antagonize Savannah authorities and disrupt an already successful protest movement.[47] It is not surprising that the Gilbert museum says little about King but focuses largely on local activists and struggles. Roger Stump suggested that streets named for King are "public symbols of community values, attitudes, and beliefs, revealing the character of both the figure commemorated and the community that has honored him."[48] As symbols, streets named for King often are contested sites that erupt through the political tensions underlying the remembrance of King along America's roadways.

The Politics of Streets Named for Martin Luther King

The symbolic work of naming streets for Martin Luther King most often takes place through highly public debate and controversy. Just the very naming of streets after King can have unforeseen consequences and can spark political opposition. In 1987 citizens in San Diego, California, and Harrisburg, Pennsylvania, voted to revoke the renaming of streets for King. Both cities later placed his name on smaller roads.[49] When the city council in Portland, Oregon, voted in 1990 to rename Union Avenue after King, more than two dozen people picketed and heckled the street-naming ceremony and more than fifty thousand people signed a petition opposing the name change. Because of this backlash, Portland voters were to be given a chance to vote on an initiative in an upcoming primary election that would have changed the name of the street back to Union Avenue, but before the election was held, a county circuit judge ruled that placing such an initiative on the ballot was illegal.[50]

The landscapes of commemorating King can serve as flashpoints around what Gary Fine called "reputational politics" and provide a mechanism for identifying and fixing political positions and for opening up political debate about King and his legacies, as well as for larger issues around race and power. The historical image of a person is a social product open to multiple and competing constructions and interpretations. There can be any number of different discourses or common ways of thinking and talking about a person and his or her contribution to society. The historical reputation of a person is used and controlled by social actors and groups who seek to advance their own commemorative agenda and divert the agendas of other parties. Fine recognized that the "control of history may be contentious, and the claims of one group may be countered by another that wishes to interpret the same … person through a different lens."[51]

Naming streets after King as reputational politics highlights any number of contemporary political issues revolving around race. These issues range from representations of King, his legacy, and his legitimacy to questions about King's resonance within African American communities and in American life more broadly. They often extend to expose basic racial and political tensions about urban economic vitality and urban apartheid, and they often work through a politics of scale to join local and national interests. The landscapes of streets named for King are politically charged and often are sites of political struggle.

Reputational politics often arise first when African Americans seek to establish the very legitimacy of commemorating King. City officials in Americus, Georgia, did not rename a portion of U.S. 19 until black community leaders planned a boycott of city businesses. Part of the controversy stemmed from the comments of a white fire official. He supported naming half of the street for King if authorities named the other half for James Earl Ray, the man convicted of assassinating the civil rights leader.[52] In Dade City, Florida, vandals painted the name "General Robert E. Lee" over nine Martin Luther King Jr. Boulevard signs, an incident symptomatic of the American South's ongoing struggles over identity and memory. In a single year, almost one hundred street signs with King's name in Hillsborough County, Florida, were either spray painted, shot at, or pulled completely from their poles.[53] Not long after officials in Mankato, Minnesota, named a small street for King, "an unidentified motorist mowed down both of … [the city's] new MLK street signs while shouting racial epithets at some passing children."[54]

In addition to legitimacy, the politics of constructing King's historical reputation through street naming is also a struggle over resonance. One of the largest obstacles facing African Americans is the prevailing assumption, particularly among whites, that King's historical relevance is limited to the black community. In Statesboro, Georgia, African Americans tried, unsuccessfully, on two occasions to have a major road identified with King. In their first attempt, black activists struggled with local veterans over naming a new perimeter highway. Veterans succeeded in representing their memorial cause as inclusive of all races and groups of people. In contrast, they depicted the commemoration of King as socially divisive, suggesting that his memory did not resonate with whites. Outspoken black leaders countered by asserting the universal importance of King's legacy. They reminded the public of his war on poverty and economic inequality—issues of great concern not only to blacks but to all Americans.[55]

In representing the street-naming issue as divisive, some whites have suggested that King—because of his legacy as a peacemaker—would not have wanted his commemoration characterized by racial conflict. For example, street-naming opponents in Chapel Hill, North Carolina, argued this point when they called on black leaders to rename a park, library, or school for King rather than the controversial Airport Road. Black supporters such as Michele Laws countered with King's own words: "The ultimate measure of a man is not where he stands in moments of comfort and convenience but where he stands in times of challenge and controversy."[56] These attempts by some whites to represent the civil rights leader's image as nonconfrontational is, according to Michael Eric Dyson, part of a larger national amnesia about King's true legacy. According to Dyson, most of America chooses to remember King as the "moral guardian of racial harmony" rather than as a radical challenger of the racial and economic order.[57] In this respect, the politics of street naming are not just about black Americans establishing the legitimacy and resonance of King's achievements but also about wrestling away control of his historical legacy from conservative whites, who have appropriated his image to maintain the status quo rather than redefine it.

In the struggle to elevate the historical reputation of King, black activists often engage in a "politics of scale."[58] On one hand, African Americans seek to extend the geographic and social reach of King's importance within cities. They suggest that his significance is not limited to the black community and hence seek to name major thoroughfares that cut across and unite different racial communities. On the other hand, opponents attempt to place spatial boundaries around the people and locations that will be associated with the commemorative naming. They interpret the civil rights leader strictly as an African American advocate and seek to confine his name to areas that do not seem to directly touch the lives and geographies of the white community. This was evident in Chattanooga when opponents argued that King's name did not belong on the western part of Ninth Street because it was no longer a "black" street and might subsequently harm white-led economic development. In resisting these efforts to segregate King's memory, street-naming proponents asserted the cross-racial legitimacy of memorializing the civil rights leader. In making the argument for renaming all of Ninth Street, U.S. representative Parren Mitchell, a former chair of the Congressional Black Caucus, explained, "All groups want monuments and symbols of their race. But with King, of course, it was more than a matter of race—it was the impact he had on this nation and this world."[59]

Many black communities see the naming of a prominent, highly visible street as a reflection of the importance that a community places on King. The naming of large, racially diverse streets allows African Americans to educate the entire community about the contributions of King. In contrast, renaming a smaller, less prominent street represents a restriction of King's image and its potential to reshape the public's historical consciousness. The rescaling of King's memory was certainly in the minds of NAACP leaders in Clearwater, Florida, when they persuaded local officials to remove King's name from a three-block stretch of road and rename a three-mile length of road that cuts through a variety of residential neighborhoods and the city's historically African American business district. As one activist contended, "If King is going to have a road named after him, it should be more significant. It should traverse different areas of the city, different boundaries."[60] Not all African American communities are successful in naming streets (or portions of streets) that reach beyond the geographic boundaries of the black community. A study in 2000 of streets named for King in the southeastern United States found them to be located in largely African American areas of cities.[61] More recent research for the nation as a whole suggests that neighborhoods intersected by these streets have a significantly larger proportion of African Americans than their respective cities.[62]

When African American activists seek to remember King on prominent thoroughfares, they often encounter harsh opposition from owners and operators of businesses along the potentially renamed street. Businesses most often cite the financial burden of changing their address as printed on stationery, advertising, and billing statements. Some opponents such as those in Zephyrhills, Florida—whose city officials voted to dedicate rather than rename a street for King in 2004—expressed fear that property values would drop as a result of being located on a street named for King.[63] There is no evidence to suggest that street naming brings a decline in property value or loss of business, as suggested by the white developer in Chattanooga. In fact, several streets named for King in this country are the focus of significant redevelopment efforts, such as those in Indianapolis, Indiana; Jersey City, New Jersey; Savannah, Georgia; Miami, Florida; and Seattle, Washington. King's memory does not necessarily cause poverty and degradation along streets. Rather, his name is often placed in poorer areas as a result of public opposition to naming more prominent places. Although resistance from business interests has significantly limited the scale of King's commemoration, large numbers of commercial establishments can be found on several streets named for the civil rights leader (e.g., Tampa, Florida; Los Angeles, California; Washington, D.C.; Portland, Oregon; and New Bern, North Carolina) (see Figure 11.6). Analyzing the almost eleven thousand nonresidential establishments in the United States that have an address on a street named for King, Matthew Mitchelson

Figure 11.6 An example of commercial development along Martin Luther King Jr. Boulevard in New Bern, North Carolina. Contrary to the prevailing vision that streets named for King are always found in poor, economically marginalized areas, the street in New Bern is the center of significant nonresidential development such as national chain stores, a shopping mall, a Wal-Mart, car dealerships, and two soft-drink bottling plants (Pepsi and Coca-Cola). (Photograph by Matthew Mitchelson, reproduced with permission.)

found these establishments to be on par with national trends in terms of annual sales volume and number of employees. However, he also found that these streets are "less industrially diverse than other places" and "host a disproportionately high number of establishments traditionally categorized as 'black businesses,' such as beauty parlors and barber shops, small retail grocery stores, and funeral parlors."[64]

In arguing against street-naming proposals, business and property owners consistently attempt to represent their opposition as not racially motivated but simply a matter of cost and convenience. And in some cases, such as in Chattanooga, whites interpret the evoking of King's image by blacks as an attempt to create racial overtones. At the same time, black proponents almost always point to this opposition as racist in nature. As pointed out by Blauner, whites and blacks often "talk past each other" because they define racism differently. According to him, "Whites locate racism in color consciousness and its absence in color blindness."[65] They tend to see anti-black racism as a thing of the past, supposedly ending with segregation, lynching, and explicit white supremacist beliefs. The African American public, according to Blauner, defines racism much more in terms of power and how certain underlying structures and institutions maintain racial oppression even in the absence of explicitly stated prejudicial attitudes. Rather than being a thing of the past, racism from the black point of view continues to exist and has taken on a much more insidious form, as evident in the unwillingness of white entrepreneurs to change their addresses.

In characterizing the racism that African Americans encounter when attempting to honor King, we might find it worthwhile to interpret business opposition to street naming as a form of "rational discrimination." According to Bobby Wilson, in his analysis of black activism and struggle in the postmodern era, the overt racism of Eugene "Bull" Connor has been replaced by a "rational discrimination" in which businesses and corporations use the pursuit of profit or cost savings as justification for not investing in African American people and places.[66] Such a position can deny the sometimes-structural impediments to equality. In addition, this "rational" form of discrimination is sometimes driven by the emotional memories of past racial tensions. For instance, in Muncie, Indiana, Ed McCloud responded to the changing of his address from Broadway to Martin Luther King by closing his appliance business of fifty years. Although McCloud expressed concern about customers not being able to find his store, he did admit vividly remembering when one of his earlier stores had been set on fire by rioters following King's assassination in 1968. McCloud added, "I swore then that I would not let the black community—or anyone else—hurt my business again."[67]

Struggles to construct the importance and meaning of King's reputation are not simply interracial but also occur within the African American community. Embracing different political goals, African American leaders sometimes disagree with each other over which street to name in honor of King. Even in Chattanooga, where black leaders formed an impressive coalition, one could find evidence of the street-naming issue being viewed in multiple and sometimes competing ways by African Americans. One particularly outspoken activist wished to see King's name on a street in a "better part of town," contending that much of Ninth Street was characterized by crime and marginal economic activity. NAACP leader George Key countered by asserting that a renamed Ninth Street "would be a symbol to let young blacks know that there is something in Chattanooga they can identity with ... to have the feeling that Chattanooga cares about its black people."[68] In Eatonton, Georgia, two African American leaders had a more visible competition—one lobbied for the naming of a major highway that ran the length of town whereas the other persuaded local officials to name a residential street within the black community. Whereas the activist advocating for the thoroughfare emphasized using King's memory to challenge and expand the historical consciousness of whites, the other activist emphasized how the naming of the residential street would focus and inspire African Americans.[69] Street naming is a negotiated process in which even black participants must balance between allowing the importance of King's commemoration to transcend the black community and keeping it within symbolic reach of African Americans.

Concluding Remarks

Like the civil rights movement they commemorate, streets named after Martin Luther King symbolize both black empowerment and struggle. Jonathan Tilove put it best when he wrote that these streets are the "geo-political synthesis of black insistence and white resistance."[70] Of course, as I have tried to demonstrate, these streets also provoke us to go beyond monolithic conceptions of "the" black community and to comprehend in fuller terms historical consciousness, geographic agency, and political activism among a diversity of African American interests. Similarly, these street-naming practices reveal that in every racialized struggle, the lines of opposition are not always or necessarily drawn along the demarcations of the black–white binary. As a rapidly growing movement that touches people in intimate and potentially controversial ways, the naming of streets for King provides a glimpse into where the country is in terms of race relations. Depending on the ultimate location that these streets take, they can symbolize the expansion of African American cultural expression and influence or simply a reentrenchment of the boundaries that have traditionally constrained black power and identity. Although named streets commemorate the civil rights movement as a completed part of the country's past, they speak, perhaps more importantly, to the still unfinished nature of King's dream of racial equality and social justice.

Notes

1. An expanded discussion of the street fight in Chattanooga and the ongoing American movement to build civil rights memorials can be found in Owen Dwyer and Derek H. Alderman, *Civil Rights Memorials and the Geography of Memory* (forthcoming). I am indebted to Ronald Foresta for first making me aware of the street-name struggle in Chattanooga.
2. Pat Wilcox, "New Request Made to Rename Ninth Street for Late Dr. King," *Chattanooga Times*, January 21, 1981, B2.
3. Pat Wilcox, "Unity Asked in Street Name Change," *Chattanooga Times*, March 25, 1981, B1.

4. Jeff Powell, "Blacks 'Rename' Ninth Street," *Chattanooga News-Free Press,* April 19, 1981, A1.
5. Pat Wilcox, "City Reverses, Renames Ninth Street for King," *Chattanooga Times,* July 15, 1981, A1.
6. Pat Wilcox, "Key Calls Change in Read House Address Racism," *Chattanooga Times,* January 16, 1982, A1.
7. Jonathan Tilove, *Along Martin Luther King: Travels on Black America's Main Street* (New York: Random House, 2003).
8. Identification of the number of streets named for Martin Luther King Jr. was carried out in summer 2003 by Matthew Mitchelson, Chris McPhilamy, and Derek Alderman by performing street-name lookups in www.mapblast.com, www.melissadata.com, and *American Business Disc* (InfoUSA). A more comprehensive data collection by Mitchelson found 777 places in the United States with streets named for King. Matthew Mitchelson, "The Economic Geography of MLK Streets" (master's thesis, East Carolina University, 2005).
9. Derek H. Alderman, "Street Names as Memorial Arenas: The Reputational Politics of Commemorating Martin Luther King Jr. in a Georgia County," *Historical Geography* 30 (2002): 99–120.
10. Hollis Towns, "Back Streets Get King's Name," *Atlanta Journal-Constitution,* October 30, 1993, A3.
11. Tilove, *Along Martin Luther King,* 5–6.
12. Quoted in Powell, "Blacks 'Rename' Ninth Street," A1.
13. Tilove, *Along Martin Luther King,* 21.
14. Joseph T. Rhea, *Race Pride and the American Identity* (Cambridge, MA: Harvard University Press, 1997).
15. Melvin Dixon, "The Black Writer's Use of Memory," in *History and Memory in African-American Culture,* ed. G. Fabre and R. O'Meally (New York: Oxford University Press, 1994), 18–27, 20.
16. Quoted in Wilcox, "City Reverses, Renames Ninth Street for King," A1.
17. Karen Till, "Staging the Past: Landscape Designs, Cultural Identity and Erinnerungspolitik at Berlin's Neue Wache," *Ecumene* 6 (1999): 251–83, 254.
18. Derek H. Alderman, "Creating a New Geography of Memory in the South: The Politics (Re)naming Streets after Martin Luther King, Jr." (Ph.D. diss., University of Georgia, 1998); Mitchelson, "The Economic Geography of MLK Streets," chap. 6.
19. Wilbur Zelinsky, *Nation into State: The Shifting Symbolic Foundations of American Nationalism* (Chapel Hill: University of North Carolina Press, 1988).
20. Roy Rosenzweig and David Thelen, *Presence of the Past: Popular Uses of History in American Life* (New York: Columbia University Press, 1998).
21. Owen J. Dwyer, "Interpreting the Civil Rights Movement: Place, Memory, and Conflict," *Professional Geographer* 52 (2000): 660–71.
22. Kenneth Foote, *Shadowed Ground: America's Landscapes of Violence and Tragedy* (Austin: University of Texas Press, 1997).
23. Tilove, *Along Martin Luther King,* 20.

24. Roger W. Stump, "Toponymic Commemoration of National Figures: The Cases of Kennedy and King," *Names* 36 (1988): 203–16.
25. "King Drive Wins City Council Approval," *Atlanta Constitution,* April 20, 1976, 1A.
26. Gary Daynes, *Making Villains, Making Heroes: Joseph R. McCarthy, Martin Luther King Jr., and the Politics of American Memory* (New York: Garland, 1997).
27. Alderman, "Creating a New Geography of Memory in the South."
28. National Public Radio, "Martin Luther King Holiday," *Morning Edition,* January 19, 1998.
29. Maoz Azaryahu, "German Reunification and the Politics of Street Names: The Case of East Berlin," *Political Geography* 16 (1997): 479–93, 480.
30. Saul B. Cohen and Nurit Kliot, "Place-names in Israel's Ideological Struggle over Administered Territories," *Annals of the Association of American Geographers* 82 (1992): 653–80.
31. Garth A. Myers, "Naming and Placing the Other: Power and the Urban Landscape in Zanzibar," *Tidschrift voor Economische en Sociale Geografie* 87 (1996): 237–46.
32. "Suggestions Submitted to Perimeter Naming Committee," Bulloch County, Georgia, April 1–June 1, 1994.
33. Donnie Simmons (member of the NAACP chapter in Bulloch County, Georgia, and street-naming activist), in discussion with the author, May 21, 1997.
34. Robert D. Bullard and Glenn S. Johnson, *Just Transportation: Dismantling Race and Class Barriers to Mobility* (Stony Creek, CT: New Society, 1997); Robert D. Bullard, Glenn S. Johnson, and Angel O. Torres, eds., *Highway Robbery: Transportation Racism and New Routes to Equity* (Cambridge, MA: Southend Press, 2004).
35. Sally Greene (city councilor in Chapel Hill, North Carolina), in discussion with the author, May 24, 2004.
36. Maoz Azaryahu, "The Power of Commemorative Street Names," *Environment and Planning D: Society and Space* 14 (1996): 311–30.
37. Quoted in Craig Pittman, "King's Fight Still in the Streets: Renaming Roads Incites Controversy," *St. Petersburg Times,* April 23, 1990, B1.
38. Mitchelson, "The Economic Geography of MLK Streets," chap. 4.
39. Andrew Charlesworth, "Contesting Places of Memory: The Case of Auschwitz," *Environment and Planning D: Society and Space* 12 (1994): 579–93; Nuala Johnson, "Cast in Stone: Monuments, Geography, and Nationalism," *Environment and Planning D: Society and Space* 13 (1995): 51–65; Owen J. Dwyer, "Location, Politics, and the Production of Civil Rights Memorial Landscapes," *Urban Geography* 23 (2002): 31–56; Jonathan I. Leib, "Separate Times, Shared Spaces: Arthur Ashe, Monument Avenue, and the Politics of Richmond, Virginia's Symbolic Landscape," *Cultural Geographies* 9 (2002): 286–313.
40. Quoted in Brenda Yarbrough, "Street Honoring King Leads to City Dump," *Atlanta Constitution,* October 30, 1992, A3.

41. Quoted in Derek H. Alderman and Owen J. Dwyer, "Putting Memory in Its Place: The Politics of Commemoration in the American South," in *World-Minds: Geographical Perspectives on 100 Problems*, ed. D.G. Janelle, B. Warf, and K. Hansen (Boston, MA: Kluwer Academic, 2004), 55–60, 57.
42. Johnson, "Cast in Stone: Monuments, Geography, and Nationalism," 51.
43. Tilove, *Along Martin Luther King*, 122.
44. "Controversy over Renaming Shames Communities," *St. Petersburg Times*, April 22, 1990, 2.
45. Gerald Ensley, "Story of a Street," *Tallahassee Democrat*, January 17, 1999, 1A.
46. Tilove, *Along Martin Luther King.*
47. Stephen G.N. Tuck, *Beyond Atlanta: The Struggle for Racial Equality in Georgia, 1940–1980* (Athens: University of Georgia Press, 2001).
48. Stump, "Toponymic Commemoration of National Figures," 215.
49. Tilove, *Along Martin Luther King.*
50. "Street's Name Switch Riles Portland Residents, Fierce Public Backlash to Avenue Named after Martin Luther King Jr.," *Seattle Times*, March 4, 1990, D5.
51. Gary A. Fine, "Reputational Entrepreneurs and the Memory of Incompetence: Melting Supporters, Partisan Warriors, and Images of President Harding," *American Journal of Sociology* 101 (1996): 1159–93, 1161–62.
52. Peter Scott, "Failure to Name Street to Honor MLK May Bring Boycott," *Atlanta Journal Constitution*, November 28, 1992, B9.
53. Pittman, "King's Fight Still in the Streets: Renaming Roads Incites Controversy," B1.
54. Tilove, *Along Martin Luther King*, 14.
55. Alderman, "Street Names as Memorial Arenas."
56. "Citizen Comments Sent to City in Reference to Renaming Airport Road," 2004, City Hall Records, Chapel Hill, North Carolina.
57. Michael Eric Dyson, *I May Not Get There with You: The True Martin Luther King, Jr.* (New York: Free Press, 2000), 6.
58. Derek H. Alderman, "Street Names and the Scaling of Memory: The Politics of Commemorating Martin Luther King, Jr. within the African-American Community," *Area* 35 (2003): 163–73.
59. Quoted in Gary Randall, "Balking on 9th Street Termed 'Asinine,'" *Chattanooga News-Free Press*, April 5, 1981, A1.
60. Quoted in Christina Headrick, "NAACP Wants Martin Luther King Jr. Avenue Moved," *St. Petersburg Times*, August 31, 2001, 7.
61. Derek H. Alderman, "A Street Fit for a King: Naming Places and Commemoration in the American South," *Professional Geographer* 52 (2000): 672–84.
62. Mitchelson, "The Economic Geography of MLK Streets," chap. 7.
63. Joseph H. Brown, "What's in a Street Name? Ask Zephyrhills," *Tampa Tribune*, May 2, 2004, 6.
64. Mitchelson, "The Economic Geography of MLK Streets," 108.
65. Brian Blauner, "Talking Past Each Other: Black and White Languages of Race," in *Race and Ethnic Conflict*, ed. F.L. Pincus and H.J. Ehrlich (Boulder, CO: Westview, 1994), 18–28, 20.

66. Bobby M. Wilson, *Race and Place in Birmingham: The Civil Rights and Neighborhood Movements* (Lanham, MD: Rowman & Littlefield, 2000).
67. Quoted in Michael McBride, "Ed's Warehouse Shutting Down," *Star Press*, October 21, 2004, www.thestarpress.com/articles/9/028224-8999-004.html.
68. Quoted in Wilcox, "Unity Asked in Street Name Change," B1.
69. Alderman, "Street Names and the Scaling of Memory."
70. Tilove, *Along Martin Luther King*, 21.

CHAPTER 12

Puptowns and Wiggly Fields: Chicago and the Racialization of Pet Love in the Twenty-First Century

HEIDI J. NAST

Would not the rights of pit bulls
Be brutally denied
If our Supreme Justices
Ruled them hazardous to our lives?
Polls of conscious pit bull cults
Strongly verify
This is the democratic question
On America's massive mind.

...

Seeking to objectively
Put the question right
MacNeil and Lehrer held an
Indepth conversation with
A southern pit bull's wife.
To enchain these noble mammals,
Mrs. Bull graciously explained,
Is like denying civil rights
To blacks and other oppressed thangs[1]

It was a hot muggy Chicago August afternoon at Foster Beach, and I had been sitting up for the past ten minutes, groggy after a nap, watching a few children overturn buckets of fine white sand and pat them into place. The small castle's creation was soothing. Out of the corner of my eye, I noticed a small terrier running over from the direction of the Park District beach house. Its body now in full view, it lifted a hind leg and urinated, inches from the children's hands. Glancing around, I saw the dog's owner in the distance, a middle-age man in shorts surveying the scene nonplussed, leash held loosely in his hand. "Hey!" I called out. "Get your dog out of here! Can't you see there are children playing! There's a leash law, you know!" The man approached a few steps and in an eastern European accent yelled back, "What's your problem? My dog is good!" Standing his ground, he looked as much pained as surprised. To him, it appears, I was harassing his beloved, which by now I had chased back in his direction.

Over the next few months, as I took regular walks by the beach, it became clear that Lake Michigan's urban edge had become a battleground of sorts between children and dogs, the terms of which varied seasonally and by time of day. Between the week prior to Memorial Day (May) and Labor Day (September), lifeguards from the Park District patrol the beach using megaphones, warning sternly those persons who have brought their dogs to the water's edge or those whose dogs run without a leash that owners will be fined and ticketed. But before 9 a.m. (before lifeguards arrive), late in the evenings, and in the unpatrolled months between Labor Day and the next Memorial Day, subtle territorial struggles manifest themselves as small children and parents are confronted by unleashed dogs. Dog owners, often looking affectionately on, are commonly surprised and sometimes even disgruntled when a child cries or is visibly upset. Even the smallest of dogs can frighten toddlers, who might be only twice as big. But the battles are not just about innocuous dogs that scare children by their size. According to Stephen Budiansky, the most prominent reason dog owners seek professional help for their dogs is aggression, which statistically is most often directed at children. According to the author, "In 1996 dog attacks cost U.S. insurance companies $250 million in claims paid out, with total costs to society estimated at $1 billion a year," a figure he noted pales in comparison to what persons in this country spend annually on dog food ($5 billion) and canine veterinary care ($7 billion).[2]

Such turf struggles are not discernible in Steve Dale's book *Doggone Chicago,* which details the city's doggie-oriented leisure opportunities.[3] In his chapter on beaches, he tells us, "Here's the deal: According to the law, dogs are not allowed on the beaches at any time. The reality is that during the off-season, the Park District allows dogs to rule the beaches. During the beach season itself, dogs can romp on the beaches before they open at 9 am and/or after they close at 9:30 pm."[4] Indeed, it seems that dogs do rule Chicago's beaches in ways that dog owners could not have dreamed of prior to the middle of the twentieth century, when beaches were abuzz largely with family sounds, when dog ownership was more limited, and when a dog's place was seen primarily to be at home. By contrast, the 1990s has seen the emergence of a certain popular assertiveness or proprietorship about a dog owner's right to assert his or her dog's public presence. In tandem with the burgeoning dog population of the United States and related commodity chains, doggy love has become "in" and normative, and many rules restricting doggie access to public venues are now seen as outdated and onerous. Dale's warning about police citations is here apropos: "Just be aware that every now and then, a police officer with apparently nothing better to do may fine users—even in the dead of winter."[5] Though similar remarks might be made of paying heed to stoplights at 3 a.m. when no cars are in sight, companionate love is not at stake, giving Dale's words poignancy.

I begin this chapter by examining the emergence of two of the most well-known doggie areas in Chicago, ones for which the city has garnered some acclaim—Doggie Beach and Wiggly Field (see Figure 12.1). I use

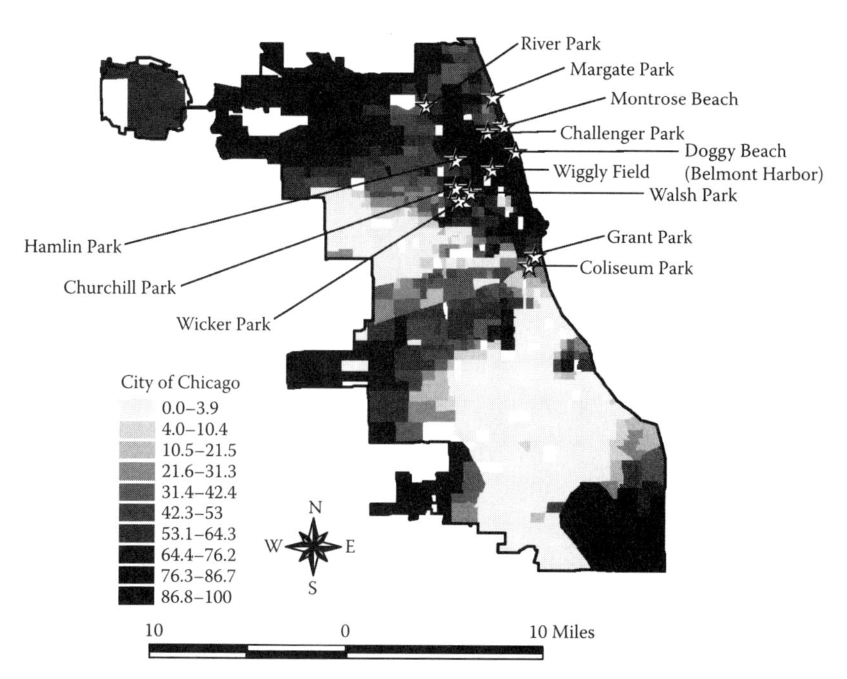

Figure 12.1 Map showing the location of two doggie beaches and eleven doggie parks in Chicago. Note how they are all located in Chicago's "white" north side. Color scale indicates the percentage of the white population. (Sources: Census, 2000; Chicago Park District, 2005. Prepared by Egan Urban Center, DePaul University.)

these locations as jumping-off points to discuss the proliferation of such landscapes in certain parts of the city and beyond and to point to how these phenomena are registering a new and popular kind of racialized love. I also discuss these places to show how it is persons with political cachet and disposable income and time who create and control these landscapes. Given that wealth in the United States is racialized, it is unsurprising that public and commercial doggie-love landscapes are located in elite white areas, often close to pet boutiques and retail outlets, a problematic elitism that is rarely remarked on.

Doggie Beach and Wiggly Field

Doggie Beach was the first beach in Chicago dedicated to dogs. Popularly seized as an effective node of doggy and owner networking in the 1990s, its emergence was tied to the previous decade's "back to the city" growth, spurred on by in-migration of elite white professionals streaming in to labor in the central business district's rapidly growing service sector. Though Doggie Beach had no official opening and has never been sanctioned by the Chicago Park District (it is technically illegal), it is Chicago's most well-known and popularly frequented beach for dogs.

In many ways Doggie Beach was a mistake: a small triangle of sand created from erosionary wave action, at its longest about one hundred meters across. Dale, in his survey of thirty-seven beaches in Chicago and surrounding suburbs, gave it a rating of three out of four bones (akin to the star system for restaurants). What Dale did not note is the importance of the cultural capital to which this beach is tied, related in turn to the city's segregated demographics: the beach is located in Chicago's north side (more than 90 percent of African Americans live in the south or west sides) at the northern end of the city's most exclusive yachting and boat harbor. The political and historical openness of this place to becoming an exercise outlet for a burgeoning dog population of new and largely white yuppie owners is evident in Dale's statement: "It took all of about two minutes for the place to grow in popularity."[6] His assertion does not explain, however, why this popularity was not evident in the more than one hundred years prior to the 1990s. After all, beaches and dogs have been around since the founding of the city. His words also do not and cannot reveal the implicitly racialized nature of this place: that here, African Americans, in particular, who make up more than 30 percent of Chicago's population, are noticeably few in numbers. African Americans, moreover, have never had the political or economic clout to establish a harbor facility on the south side or the cultural and political cachet required to otherwise appropriate illegally a piece of city turf for their dogs.

Figure 12.2 Wiggly Field in Lincoln Park, August 2005 (photo by Mona Aburmishan).

A similar political economic context and similar sorts of racialized class politics produced Wiggly Field a few years later (see Figure 12.2). Located two miles south of Doggie Beach and about three miles north of the central business district, it got its name from nearby Wrigley Field, home of the Chicago Cubs. Opened to the public in 1997, Wiggly Field was the first doggie park that the city sanctioned, its space carved out of the Park District's Grace-Noethling Park. Rimmed by a chain-link fence, its small square-shaped grounds were laid out simply.

Like Doggie Beach, Wiggly Field was sited in the exclusive, largely white neighborhood of Lincoln Park. Like most redlined urban areas abandoned after World War II, Lincoln Park had been a ghetto throughout the 1960s, its most notable residents being Puerto Rican. With the end of industrialization in the 1970s and the creation of an elite postindustrial service sector, areas near and within downtown (the "loop") were gentrified and taken over by a largely white populace who desired handsome, large, vintage homes. Today, Lincoln Park is one of the most highly priced and prestigious residential areas in the city and is accordingly demographically limited.

Dale noted that Wiggly Field "came about because of the antidog campaign launched at several Near North Side parks, including Oz and Jonquil Parks. If those antidog terrorists were to get their way, where would the dogs go? After several years … a committed legion of dog lovers finally persuaded the Park District to devote a park for dogs."[7] This committed

Figure 12.3 Wiggly Field water fountains were created at two levels: one for dogs, the other for humans, August 2005 (photo by Mona Aburmishan).

legion formed itself into the People United to Preserve the Parks (for pets), an association now dedicated to maintaining the park. Thanks to their efforts, the formerly simply configured park was transformed into a landscaped canine exercise area and playground, with two water fountains—one for humans and the other at near-ground level for dogs (see Figure 12.3).

The doggie fervor and racialized wealth and politics involved in creating Doggie Beach and Wiggly Field in the mid- to late 1990s is in sync with what has been observed nationally. In West Chester, Ohio, about twenty-five miles north of Cincinnati, near an exclusive and largely white community, a similarly named Wiggly Field dog park was created in 2002. Although initial funding came from a single resident and a major pet corporation, it is now managed by West Chester Parks and Recreation. As the official website boasted in a June 2002 announcement, "The three-acre dog park is part of the overall master plan for the Voice of America Park and is now open! Iams Wiggly Field is one of the area's premier canine facilities with a large fenced grass area for dogs to play and explore, dog drinking fountains, benches, paved staging area and a secure gate. While dogs are permitted throughout the Voice of America Park on leash, Wiggly Field will provide an opportunity for well-behaved dogs to run unleashed." In addition to opportunities for the public to donate trees and park benches to embellish the doggie landscape, smaller monetary donations are also accepted. For $75, for example, any member of the public can purchase one of two thousand canine commemorative bricks that are to be laid out "in a bone shape creating a waiting area where pet owners may congregate while their dogs are enjoying the grassy fenced-in area." The bricks are meant to memorialize dogs who have died, each brick inscribed with the name of the beloved dog; more than one name can be accommodated, provided the names do not exceed three lines of type.[8] The doggie-focused website advertising the park and its commemorative bricks features cursors consisting of myriad little dog paws, an aesthetic and iconic device common to many doggie-oriented websites and pet-oriented merchandising logos.

Propagation

During the past decade, doggie love has irrupted into all kinds of social landscapes, services, commodity chains, and imaginaries. Doggie love's material expressions and geographical reach now trenchantly cross national landscapes, burrowing into the racialized fabric of everyday local lives. As a result of Chicago's ever-enlarging, politicized doggy circles, the city now ranks as one of the best doggy-tourist destinations in the nation, the creation of doggy-related surveys being a peculiar invention of the turn of the twenty-first century. In 2005, for example, the third annual survey of the website www.dogfriendly.com of the top ten most dog-friendly cities in Canada and the United States ranked Chicago number one, up from third place in 2004 and fourth place in 2003.[9] This website, like dozens of similarly doggie-oriented websites, was created in the late 1990s (it was launched in 1998) and boasts on its "About Us" link that its parent organization, DogFriendly.com, Inc., "is the leading provider of nationwide city guides and travel guides for dog owners … [and is] dedicated to finding places that people and dogs can enjoy together. If you want to travel, sightsee, or just go around town with your pooch, then you've come to the right place! We do not believe that a well-behaved dog should be discriminated against and therefore *we focus on listing only places that allow dogs of ALL sizes and breeds*" (emphasis in original).

The structure and content of the site and its web links show that its audience is largely white. Of the 582 posted photographs of viewers' dogs (on Santa's lap, on the deck of a speed boat, in swimming pools, in dog shows, on sofas, etc.), for instance, the few that featured humans—less than 6 percent—showed all of them to be white, except for a South Asian woman located in New Delhi. (The demographic breakdown is eighteen women, eight men, four girls, two boys, and one infant.)

Figure 12.4 The fence between the beaches for humans and for dogs at Doggy (Montrose) Beach. Gimme Paw Café is located to the left, just out of view, August 2005 (photos by Mona Aburmishan).

Given the groundswell of this love and its racialized dimensions, it is no surprise that doggie forces on Chicago's white north side gained enough political leverage to lobby successfully for the first-ever legitimate doggie beach, christened officially by the Chicago Park District in 2002 (see Figure 12.4). The beach comprises the northernmost reach of Montrose Beach, two blocks from my home in Uptown—another formerly redlined ghetto now being reclaimed by a largely white middle and upper-middle class. With largely white gentrifiers moving into Uptown and largely poorer African Americans, Latinos, and Native Americans moving out (most of the neighborhood's housing is being converted to upscale condominiums), Uptown appears deceptively diverse.

The creation of the doggie landscape at Montrose Beach was not uncontentious. The main political complainants were bird lovers and conservationists, rather than parents per se. On June 12, 2002, bird lovers had mobilized to confront MONDO—Montrose Beach Dog Owners (since renamed MONDOG)—at a public meeting where the latter planned to table a proposal for the dog beach. Held at the Lincoln Park Cultural Center and attended by the alderwoman of the Montrose Beach district, Helen Schiller, the complainants argued against the sponsorship MONDO sought from the Lincoln Park Advisory Council (which advises on the entirety of beaches rimming the lake, collectively known as Lincoln Park—not to be confused with Lincoln Park neighborhood). Their main argument was that "there is a natural prey-predator relationship between dogs and shorebirds, which nearly always fly away at the approach of a dog. After several disturbances, many simply leave Montrose Beach, underfed and unrested, with the success of their migration and perhaps survival at risk. Dogs and their owners can go elsewhere in Chicago: Shorebirds can only use the water's edge and will only use Montrose Beach on Chicago's north side."[10] Like the MONDO representatives, the opposition was privileged.

Despite opposition, the doggie beach opened, immediately becoming a canine haven cordoned off from the main beach by a chain-link fence. Its state-sanctioned status has meant that the beach holds political legitimacy, unlike its illegitimate cousin, Doggie Beach, permitting its use for formal public occasions. One of the most high profile of its doggie uses had to do with the launching of *Benji: Off the Leash!* released in August 2004. Joe Camp, the film's director and Benji's trainer, conducted a nationwide search for the film's doggie star, observing more than one thousand dogs for the part. One of the finalists was Shaggy, a male mutt that Camp adopted from a Chicago animal shelter. Though Shaggy did not win the role of Benji, he was apparently charmingly goofy enough for Camp to rewrite the film's script to include him in as the main character's comic sidekick. Thus *Benji: Off the Leash!* became a mainstream doggie film featuring real dogs engaged in a comedic partnership, their relationship propelling the plot line. Camp, Benji, and Shaggy were subsequently invited on June 19, 2004, to kick off the film's premiere party at Montrose Beach (proceeds to benefit Shaggy's former animal-shelter home). The site additionally was used to host Chicago's second annual "Bow-Wow Beach Bash." All this happened only two years after MONDOG's contentious June 2002 meeting.

Throughout his book, Dale posts asides about there being few dog–human tensions in the places he visited that he could discern. Perhaps this has to do with the fact that he visited, interviewed, and spoke with folks in the presence of his dogs, in which case few persons might feel the chutzpah to complain. By contrast, when I petlessly inquired into how the boundaries of Montrose's doggie beach were maintained, lifeguards groaned about how owners regularly ignored the leash law before reaching the beach, about dog owners becoming combative when told that their dogs could not have the run of the beach, about how tiring it was to keep reminding people not to keep dogs on the humans' side of the beach, and about the verbal abuse or ridicule they often endured from dog owners upset over city restrictions. These proprietary feelings of doggie owners register the familial ways that many postindustrially situated humans have made libidinous their relationship to their pets. Yet these familial feelings are registered in markedly racialized ways.

The racialized ways of pet love are poignantly evident in the coincident creation of Uptown's "Puptown," a small wrought-iron-fenced, gate-enclosed canine exercise area at the corner of my block, a stone's throw from the Lawrence bridge under which the homeless sleep and steps from the adjacent park's benches, where the homeless congregate. In 2001 at our overwhelmingly white block club meetings, the design decisions about Puptown figured largely. Block clubs are a uniquely Chicago institution: blocks of neighbors get together to discuss issues of interest to them, block club leaders becoming links to their city ward representatives (alderwomen and aldermen). It is through these clubs that area improvements take place, especially when residents' incomes provide a tax base that can fund those improvements and when residents have the leisure time and cultural political capital needed to make their voices heard.

In 2000 the Puptown Advisory Council (PAC) formed, its members hailing largely from Uptown. Yet its website notes that the PAC is a "Margate Park dog owners' group," a more socially restrictive appellation. Although Margate Park is indeed the name of the nearby Park District house, it is also the name real estate agents give to the most elite residential section of Uptown, what residents of the area proudly recall as having been the pre-Depression-era Hollywood of the United States. On Castlewood Street, for instance, many mansions of former movie stars remain. Unlike other areas of Uptown that became populated largely by the poor, this one street remained white and wealthy while most other streets converted into high-density rental housing. Designating park's affiliation to Margate Park hence distances it from associations with the larger area of Uptown, which the park supposedly serves. At the same time, the name "Puptown" dampens the historical notoriety and racialized poverty of the area, inoculating its users to the area's human needs and past.

Some of the worries on which the PAC centered early on involved choosing a doggie-appropriate water fountain (set near ground level) and the kind of surface most appropriate to pups' feet (asphalt was seen to be the most sanitary) and figuring out how to get community support to maintain the doggie park's grounds and to make sure that doggie and owner courtesy rules were followed. The PAC's unofficial website, begun in 2000, relays that its members comprise "a group of concerned dog owners who wish to work with the Park District, the police department, and other dog owners to improve conditions for all dogs (and humans) who use the Margate Park dog-friendly enclosed area on Lawrence Avenue at Marine Drive."[11] Meeting the first Monday of every month in the Margate Park Fieldhouse, this very industrious group worked hard on behalf of their dogs' lives. On October 29, 2000, for instance, they hosted a Doggie Halloween Costume Party: "The party starts at 2 pm; costume judging begins at 2:30. Don't miss this great chance to meet your north-side dog neighbors, win prizes (including a raffle—no costume required), and check out the latest canine costumes." As in the days of Catholic parish events, there was a community raffle with corporate goodies donated. "Raffle items include merchandise from Barker & Meowsky. All dogs will receive a special treat bag. $5 donation requested to benefit Puptown's drinking fountain and maintenance fund! Write for more information or to volunteer to [*sic*] lobbying for doggie-park-appropriate lighting." Most human visitors to the park, although hailing from many parts of the city, are white.

As in the case of Wiggly Field's namesake park in Ohio, Uptown's Puptown has an unintended namesake in San Diego: the Puptown Doggy DayCare, a temperature-controlled day care facility sanitized daily that features rubber flooring inside ("for bruise-free tumbling") and "supervised free play, socialization, command reinforcement (positive only), and tons of TLC in a safe, secure, fun environment." Just six blocks east of the San Diego Padres' Petco Park, it features "outdoor space for fresh air, sunshine, and potty breaks" along with "fresh distilled water" and "lots of comfy beds for our canine clients to nap on." Its location, like those of similar facilities in Chicago, is situated on prime real estate (downtown, near the oceanfront), making its free private parking especially valuable. Like any child day care, owners must call or e-mail ahead of time to schedule services, including crate-free overnight "doggy stays." Here, "Your dog can play all day, enjoy dinner here, frolic all evening, then snooze through the night."[12]

The proximity of Puptown Doggy DayCare to Petco Park speaks to the pervasiveness of the socioeconomic and political networks being fashioned through doggie love's landscapes and practices. Political momentum to build the massive baseball-only stadium (officially opened in 2004) increased in 1998 after the Padres made it into the World Series, a time coinciding with a meteoric rise in major pet corporations' profits. Following an initial construction hiatus in October 2000, stadium building began anew in November 2001 after the City of San Diego approved a $166 million bond order. "In January 2003, Petco Animal Supplies, Inc. purchased the naming rights for $60 million over 22 years, thus the ballpark was named Petco Park."[13] The magnitude of the contribution signals the economic vitality of doggie love and the economic forcefulness it can now command.

California in many ways pioneered the racialized and class-inflected accommodations described here. In 1972, for example, San Diego opened Dog Beach, perhaps the first state-sanctioned doggie beach in the nation, "a legal off-leash paradise for dogs ... [offering] 38 acres of sand where the San Diego River meets the Pacific Ocean." Like the intensification of doggie-oriented commodities and services that occurred nationwide thereafter, Dog Beach became a local anchor for doggie-oriented investments, including the "self-serve dog wash" two blocks away, Dog Beach Dog Wash, opened in 1993.[14]

Conclusions

Today, Chicago boasts eleven dog parks, with several others in the planning stages. All of them, save one, are located in gentrified areas in the city's white north side. The only south side park is located in the elite enclave of Hyde Park, a place famous for having removed its poorest residents (African Americans) through pioneering the historically and overtly racist national urban renewal policy.

In this chapter I have tried to show how a very particular kind of pet love, one highly commodified and tied to doggie-dedicated landscapes, is racialized and wealth inflected. Although technically any person can visit and enjoy the public spaces discussed here, their racialized location and privileged creation cannot be overlooked. This is not to say that the poor do not own pets or go to these places but to say that there are racialized class dimensions to pet ownership and the ability to buy into the new material expressions and media of pet love. Dogs figure largely, in part because they are the most social of domesticated animals and thus amenable to (or tolerating of) human excess. They also can be cheap and even free, allowing persons of any economic means to procure a pet. Yet racialized class dimensions are evident in where pets live, their material circumstances, and how they are used. Dogs bred for gaming fights or used in poor homes in dangerous neighborhoods are not likely to be dressed in faux biker jackets or to have a luxury bed. Conversely, guard dogs used in wealthy and poor neighborhoods are often deployed differently (one guards wealth whereas the other might guard against danger) and mean different things (a pure bred of American Kennel Club status versus a mutt). I do not document these many nuances here but rather sketch out the rough contours of pet love's creation and incorporation into hegemonic social relations primarily since the 1990s.

Notes

1. Excerpt from Michael Warr, "Pit Bulls in America—Or the Depth of Our Democracy," in *We Are All the Black Boy* (Chicago: Tia Chucha, 1990).
2. Stephen Budiansky, "The Truth about Dogs," *Atlantic Monthly,* July 1999.
3. Steve Dale, *Doggone Chicago* (Chicago: NTC Contemporary Books, 1999).
4. Ibid., 24.
5. Ibid., 25.
6. Ibid., 30.
7. Ibid., 9.
8. See http://www.westchesteroh.org/parksandrec/wiggly/ (accessed August 2004); see also http://www.westchesteroh.org/parksandrec/parks/#voa for a general description of the park.
9. See http://www.dogfriendly.com/server/newsletters/toptenc2005.shtml.
10. See http://www.ece.iit.edu/pipermail/bcnnet/2002-June/000169.html.
11. See http://www.publica.com/puptown.html.
12. See http://www.puptowndoggydaycare.com/pages/725136/index.htm (accessed August 2004); see also http://www.californiafind.com/Localities/S/San_Diego/Business_and_Economy/Animals/Puptown_Doggy_Day-Care_L46374.html.
13. See http://www.ballparksofbaseball.com/nl/PetcoPark.htm.
14. See http://www.dogwash.com.

Contributors

Derek H. Alderman is an associate professor of geography at East Carolina University and a coeditor of the peer-reviewed journal *Southeastern Geographer.* His research interests include geographies of public commemoration, the politics of place naming, and the cultural and historical geography of the American South.

Daniel D. Arreola is a professor in the department of geography, and an affiliate faculty with the Center for Latin American Studies and with the Hispanic Research Center at Arizona State University. He has published extensively on topics relating to the cultural geography of the Mexican American borderlands. He is a coauthor of *The Mexican Border Cities: Landscape Anatomy and Place Personality* (University of Arizona Press, 1993), author of *Tejano South Texas: A Mexican American Cultural Province* (University of Texas Press, 2002), and editor and author of *Hispanic Spaces, Latino Places: Community and Cultural Diversity in Contemporary America* (University of Texas Press, 2004). He has lived and taught in three of the four American states that line the U.S.–Mexico border.

Michael Crutcher is an assistant professor of geography and African American Studies at the University of Kentucky. His interests include urban performance traditions, cultural tourism, and Southern black identity. He is presently completing a book for LSU Press titled *Tremé Daze: The Creation, Destruction, and Commodification of a Downtown New Orleans Cultural Landscape.*

Samuel F. Dennis Jr. is an assistant professor in the department of landscape architecture and the Nelson Institute for Environmental Studies at the University of Wisconsin, Madison. He is both a cultural geographer and a registered landscape architect, and his primary interests lie in the social meaning of public spaces and, in particular, the intersection of place making and the social construction of identity.

James Duncan teaches cultural geography at the University of Cambridge and is a fellow of Emmanuel College. His publications include *The City as Text: The Politics of Landscape Interpretation in the Kandyan Kingdom* (Cambridge University Press, 1990) and *Landscapes of Privilege: The Politics of the Aesthetic in Suburban New York* (Routledge, 2003). He is coeditor of *A Companion to Cultural Geography* (Blackwell, 2004).

Nancy Duncan teaches cultural geography at the University of Cambridge, where she has been a fellow of Fitzwilliam College. Her publications include *BodySpace: Destabilizing Geographies of Gender and Sexuality* (Routledge, 1996) and *Landscapes of Privilege: The Politics of the Aesthetic in Suburban New York* (Routledge, 2003).

Dianne Harris is an associate professor of landscape architecture and architecture at the University of Illinois–Urbana/Champaign, where she also holds affiliate appointments in the departments of history and art history. She is coeditor (with Mirka Benes) of *Villas and Gardens in Early Modern Italy and France* (Cambridge University Press, 2001), and the author of *The Nature of Authority: Villa Culture, Landscape, and Representation in Eighteenth-Century Lombardy* (Pennsylvania State University Press, 2003) and *Maybeck's Landscapes: Drawing in Nature* (William Stout, 2005). She has an edited volume in press titled *Sites Unseen: Essays on Landscape and Vision*. She is currently guest-editing an issue of the *Landscape Journal* devoted to "Race and Space," and she is writing a book that examines ordinary postwar houses and gardens in the United States (1945–1960) as frameworks for assimilation and the reinforcement of racial constructs and class assignment.

Steven Hoelscher is a cultural geographer with research interests in American landscape, ethnicity and race, and U.S. cities. His books include *Heritage on Stage* (University of Wisconsin Press, 1998) and *Textures of Place* (ed.) (University of Minnesota Press, 2001), and he has published articles in the *American Indian Culture and Research Journal, Annals of the Association of American Geographers, American Quarterly,* the *Journal of Historical Geography, Ecumene,* and *Geographical Review.* He lives in

Austin, where he is an associate professor of American studies and geography at the University of Texas.

Gareth Hoskins is a lecturer in human geography at the University of Wales, Aberystwyth. He teaches cultural, historical, and urban geography, and his research interests revolve around geographies of race and immigration and the politics of scale, narrative, and material culture. His work on Chinese immigration and Angel Island can be found in the *Journal of Historical Geography* and *Environment and Planning (A).*

Jonathan Leib is an associate professor in the department of geography at Florida State University. His research interests include political geography, race and ethnicity, and cultural geography, with an emphasis on political and cultural change in the American South. His research focuses on issues of race, redistricting, voting rights, and the politics of representation on the contemporary southern landscape involving Confederate and civil rights iconography.

Heidi J. Nast is a cultural geographer interested in studying the importance of childbearing in various political and economic contexts. She is the author of *Concubines and Power: Five Hundred Years in a Northern Nigerian Palace* (University of Minnesota Press, 2005).

James Rojas holds a master's of city planning and a master's of science of architecture studies from the Massachusetts Institute of Science. His research is one of the few studies on U.S. Latino built environments. It has been widely cited, and excerpts have been printed in publications such as *Places* and the *New York Times.* For the past fifteen years, he has lectured extensively on his research at universities, colleges, conferences, high schools, and community meetings.

Richard H. Schein teaches cultural and historical geography at the University of Kentucky, where he holds a joint appointment in the College of Design and is a member of the Committee on Social Theory. He is coeditor of *A Companion to Cultural Geography* (Blackwell, 2004). He thinks a lot about cultural landscapes.

Index